SEVEN DAY STUDY PLAN

	SESSION 1	SESSION 2	SESSION 3
DAY 1			
DAY 2			
DAY 3			
DAY 4			
DAY 5			
DAY 6			
DAY 7			

Vineeth Remanan

BASIC TRIGONOMETRY IN ONE WEEK

International Study Version

Learn

Something

New

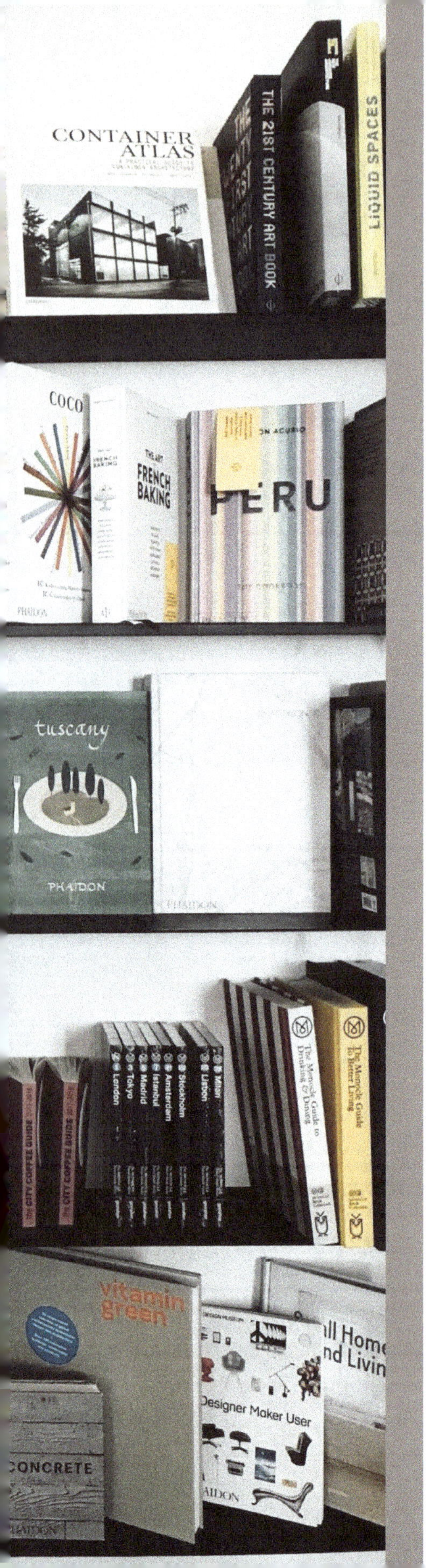

BASIC TRIGONOMETRY IN ONE WEEK

ISBN: 978-93-5526-889-1

www.neolearningbooks.com

Mathematics – Trigonometry

Written, edited and designed by
Vineeth Remanan.

Email: learnsomethingnew@neolearningbooks.com

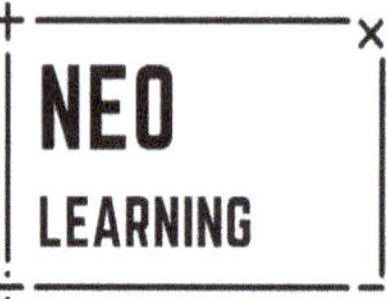

Learn
Something
New

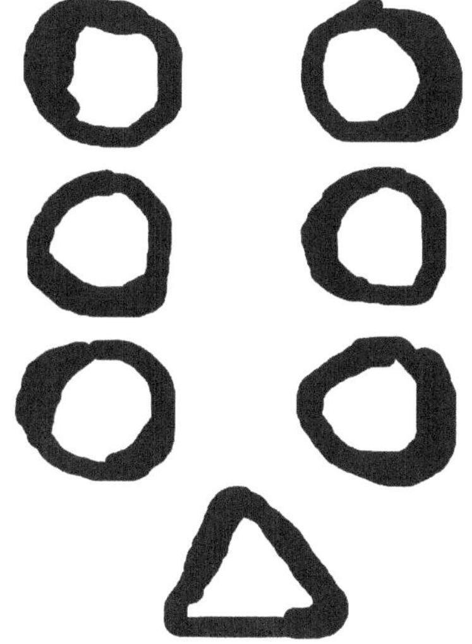

CONTENTS

MONSTER
AND HORROR
MOVIES

How to use this book - A preface on learning

Learning is one's wilful attempt to acquire knowledge at the expense of time and intense focus. It requires the desire to understand, the resolve to push through, the patience to persist and the courage to fail.

No knowledge in this world is incomprehensible, no skill in this world in unlearnable. After all, it was mere mortals like us who invented them.

The only hard part is the path one has to follow. A path of dedication, discipline, consistency and commitment. That is the only prerequisite required to start this course.

About this book

This book is a short yet consistent introduction to the world of trigonometry. It is divided into seven chapters – "The Seven Days", each consisting of three sessions, making it a total of 21 sessions, to be completed in one week.

But there are some caveats.

Identify yourself

Before starting the course, find out what kind of a learner you are?
Are you a fast learner, or a slow learner? Are you good at math, or having a hard time with it?

Use this book as a one week course, only if you are a fast learner. Otherwise, don't.

- **If you are a fast learner**:

Think of this as an athletic training camp. Athletes and competitive sports persons often participate in organised and rigorously focused training schedules to prepare themselves for upcoming sporting events.

For instance, an intense MMA training camp usually consists of 4-5 hours of training per day, split into 2 or even 3 sessions, for 5-6 days a week.

Constructing a similar schedule involves splitting a day into three sessions of 2 hours each, with enough time gaps to reset. The resetting involves refreshing, snacking, meditating and recalling the topics of the previous sessions.

Read through the topics and work out the problems. Continue to the next session only if you are confident about the current. Even if you are fast, sometimes you may not be able to keep up. Don't worry, that's ok. Take your own time. Forget the time constraints, and focus on learning.

- **If you are a slow learner**:

Appreciate the fact that every individual is unique and different. Thomas Sowell once wrote, "Nobody is equal to anybody. Even the same man is not equal to himself on different days." The same principle also applies to learning.

While fast learners readily solve problems, slow learners acquire a deeper understanding of the subject. They are not fast, because they need to convince themselves the things they are studying. Once they do that, they remember things for a longer period of time. For them, time constraints are pointless.

Unfortunately, the world is designed for quick learners, who often excel in exams and achieve higher GPAs. As a result, the slow learners whom I call the deep learners, are often marginalized and are treated as insignificant. But these are the people who might "cure our cancer or build multimillion-dollar industries".

So, if you feel left out, don't worry. Move at your own pace, take your own time, work hard and push it to the limits. The only thing that matters is the willingness to learn and the belief in yourself.

Principles of Brain-Based Learning (BBL)

The human brain is a sense making machine. Its purpose is to make sense of the information coming in through our senses, and construct various logically coherent and ever-changing structures of data. These data structures are manifested physically in the form of nerve cell connections called synapses.

One cubic millimetre of the cerebral cortex contains roughly 300 million synapses! The total number of connections in the entire brain is literally innumerable. They are responsible for our bodily functions, memory, emotions and our subjective free will.

When we learn a new subject or a skill, the nerve cells realign themselves forming new connections in order to accommodate the new information. This is called neuroplasticity.

But the new connections may not be strong as the existing ones, meaning, they may not conduct information quickly and effectively like the others. But the more we use them, they become stronger, faster, and more efficient. This is why we remember our friend's name, not our friend's friend's.

A new word, a new skill, or a new name is committed to our memory only if we use it often. Otherwise, the connections become weaker and weaker, ultimately resulting in oblivion.

This brings us to the first practice of active learning.

Spaced Repetition

The spacing effect is considered as "one of the most robust effects in psychology". It is an evidence-based learning technique in which the newly formed connections are repeatedly activated with enough time gaps in between.

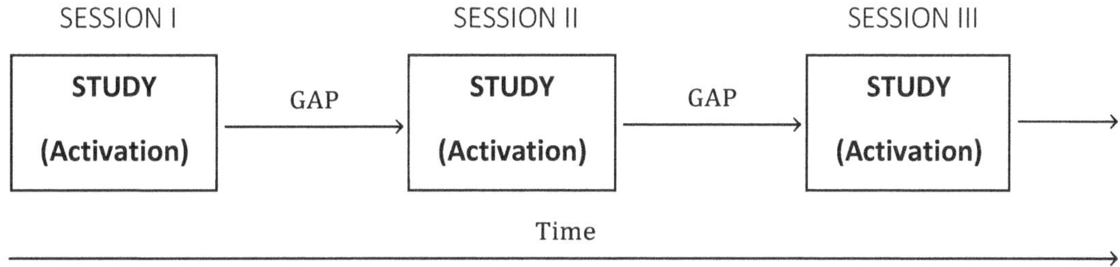

By the repeated activation, our brain recognises the importance of those connections, forcing it to strengthen the associated synapses. But the human body is not capable for a long and continuous expenditure of energy, without some form of rest. So, the gaps between the sessions are also important. A good night sleep, a daytime nap, or even a few moments of quite wakefulness could significantly improve our learning and memory.

So, instead of long, continuous study sessions, break up and distribute your learning throughout your day, or your week, or your month, with enough gaps to rejuvenate.

A few printable resources for planning and scheduling are available on our website for free. See back cover for details.

The Learning Process

The spaced repetition is shown to yield remarkable results through the repeated activation of the synapses in both clinical and academic settings. But a mere statement that: 'Studying with enough time gaps is good' is nothing more than a prescription for a scheduled study routine.

It doesn't answer about the nature of the learning process, or the methods involved in the consolidation of knowledge, neither an effective retrieval strategy to recover the stored information. For that, a further probing into the nature of learning and memory is essential.

Consider the scenario in which we are starting to 'learn something new'. Regarding the high success rate of the spaced repetition, we intend to follow such a routine.

On the first session, don't expect to learn everything. If you were able to, that's ok, good job. If you couldn't learn anything, don't worry. No one is expecting you to become a master in half an hour. Another outcome may arise in which you might tackle some concepts, but some were confusing and unclear.

If you have comprehended everything, start the next session by recalling them, then move on to the next.

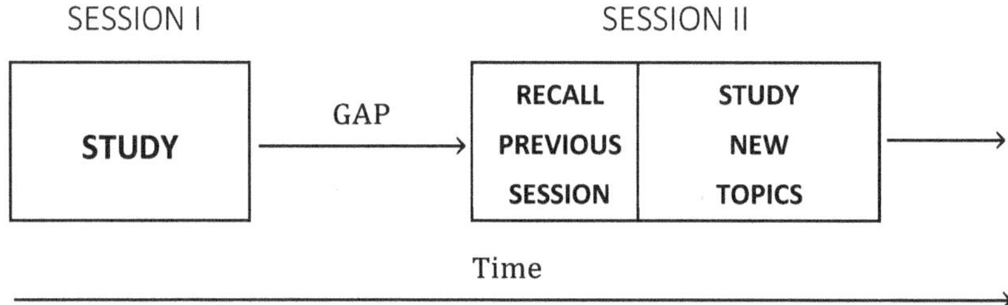

If you couldn't comprehend anything, don't worry. Remember, learning is never easy, even for the greats. So, be patient, study the topics once again and repeat until comprehension.

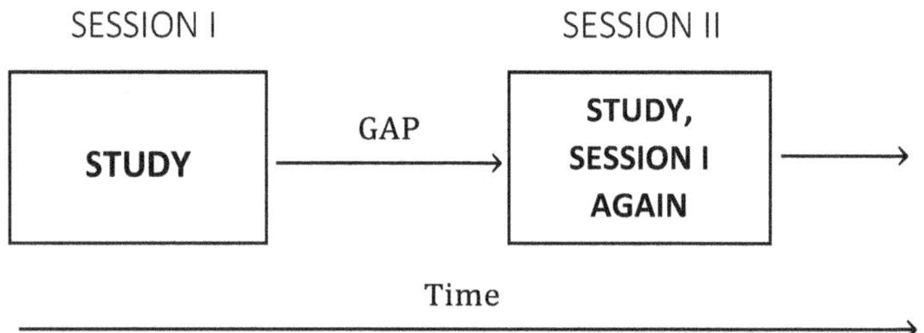

If you find yourself in the third scenario, that is you've tackled some concepts, but some were unclear. Start by recalling the concepts that you've comprehended, then study the topics that were unclear.

For instance, out of the topics A and B, assume you have learned only A. Then on the next session, recall A, then study B.

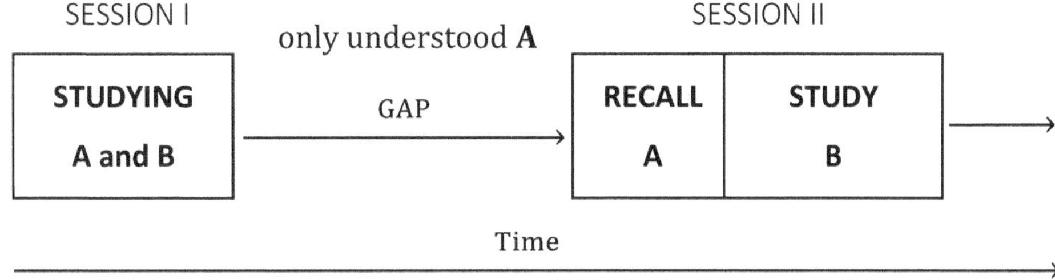

Flash Card Recalling (An Effective Retrieval Strategy)

Even though we have used the word 'recalling' quite a few times, we haven't truly discussed what it actually is.

Recalling is simply pulling information straight out of your brain without study notes or reading materials.

The absence of reading or watching forces our brain to simulate the activity originally created during the learning process. It is the most effective way to consolidate a long-term memory.

An effective execution of this technique can be done using flash cards. They are small, concise paper cards bearing keys to a particular topic. While recalling, the student must read the key and retrieve the answer.

For instance, let's say we have learned that the first president of the United States was George Washington. Then at the end of the session, make the flash card.

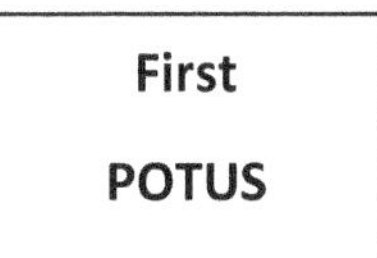

POTUS meaning, President Of The US.

On the next session, instead of reading the notes, just see the card and retrieve the answer.

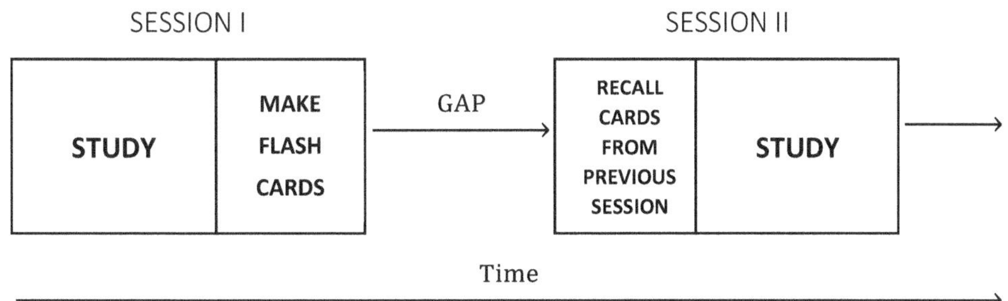

A key can be anything but the answer. It can be a code, a question, an acronym or a phrase. Don't write multiple keys into a single card. A single card should be for a single key, and always recall one at a time.

Focus – The prerequiste for learning

Up until now, all of our discussions were concerned with the process of learning and the methods involved in it. But we haven't quite discussed what its prerequisites are. Of course, an emotional motivation is required for learning. But emotions are also the universal prerequisite for all 'human actions.'

The thing that differentiates learning from almost anything, is that a focused mind is a necessity.

But in a world full of distractions, the ability to focus is almost a rarity. It is a superpower to narrow down our attention to a sole subject. It is also a test of our minds and ourselves to achieve such a state.

Recent studies in the field of learning concludes that slow learning along with some learning disabilities might be connected to the inability to focus. Therefore, it shall be the duty of every learner to prioritize the nurturing of a sound and a focused mindset.

A regular practice of meditation might seem silly, but is shown to improve our focus, attention, perception, memory, language, and even the control over our emotions.

Major corporations like Apple. Google, Nike etc. promote mediation amongst their employees in order to increase their productive output.

The benefits of meditation to mental health and productivity are enormous. It is also an ongoing and an exciting field of contemporary research.

Steps for meditation

i. **Be comfortable:** There is no fixed position or posture for meditation. You can either be on a chair, or on a bed. You can be in your room, or on a bus. As long as you are comfortable, it doesn't matter where you are.

ii. **Breathe in, Breathe out:** Meditation is all about being in the present. At present, what are we doing?... We are breathing.

So, close your eyes, breathe in, breathe out. Don't try…. just feel.

Feel the air rushing in and out. You can either focus on your nose, or on your chest, or on your belly.

iii. **Lost in thoughts:** The moment you start breathing, your mind starts to wander. Various thoughts rush into you, grabbing away your attention, and the next thing you know, you are completely lost in thoughts.

Don't be worried and don't get frustrated. It is our very nature to produce thoughts. The key to meditation is not to avoid them, but to notice them. Notice that your mind has wandered. If you have done that, you have unlocked it.

iv. **The comeback:** After noticing your wandered mind, the next logical step is to grab it, place it on your breath and start over once again.

Meditation is not a linear process, it is cyclical. You focus on your breath, you get distracted, you notice your distraction, then comeback and repeat the process over and over until you are relaxed, calm, confident and ready to learn.

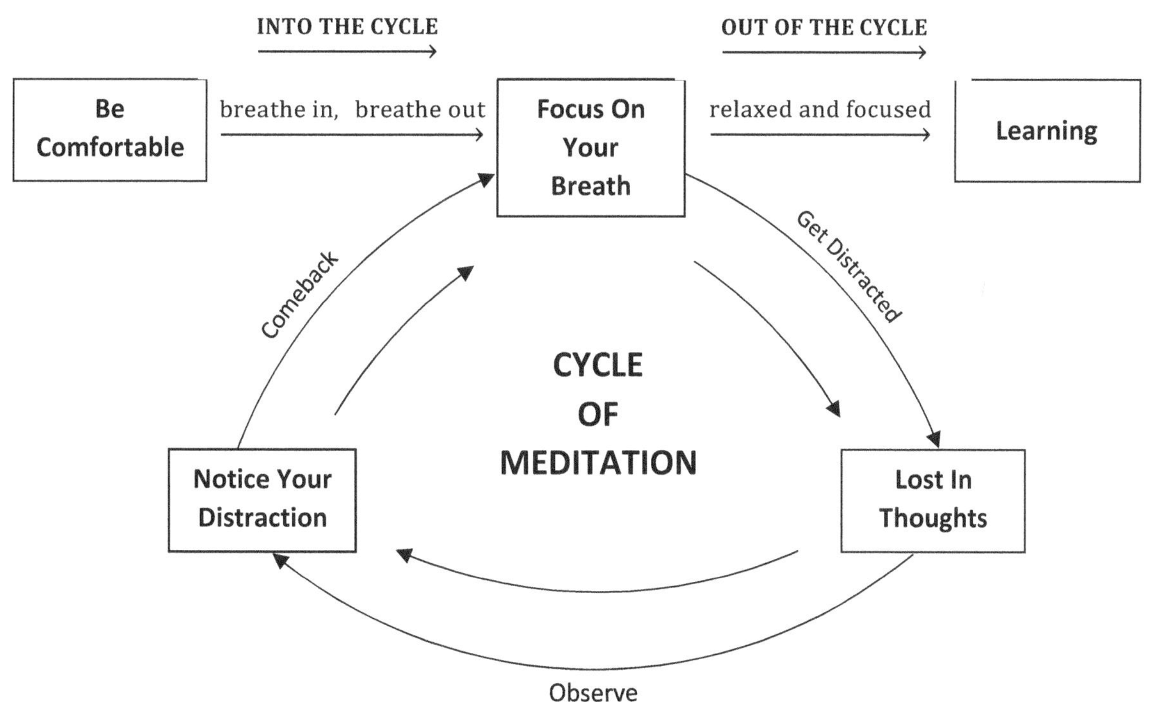

The fulfilment and satisfaction involved in a deep and meaningful learning is almost forgotten in these modern times. True learning is misplaced with cram courses and true satisfaction is misunderstood as GPAs. Even though this book is intended for a short course in mathematics, it shall not deceive the reader in believing that, there is nothing more to it. The reader should see this course as a gateway, a gateway into a wider, richer and a more profound realm of knowledge.

Good luck

Vineeth Remanan

Neo Learning

Day One

Session One
Length and angles

Radian v/s Degree

Session Two
Chopping up a degree

Counting seconds

Session Three
What is Trigonometry

Trigonometric Functions

Plan your work for today and every day, then work your plan

\- Margaret Thatcher

SESSION ONE

LENGTHS AND ANGLES

Let's get started. Lengths and angles are the two most important quantities in all of geometry. Almost all concepts in geometry when simplified are just relations among lengths and angles.

Ok, then what are those?

Length is the extend of something from end-to-end. For instance, the end-to-end separation of your hands, or the distance you walked across your room (see figure 1). Length is commonly measured in units of meter or foot, and is often denoted by English small letters.

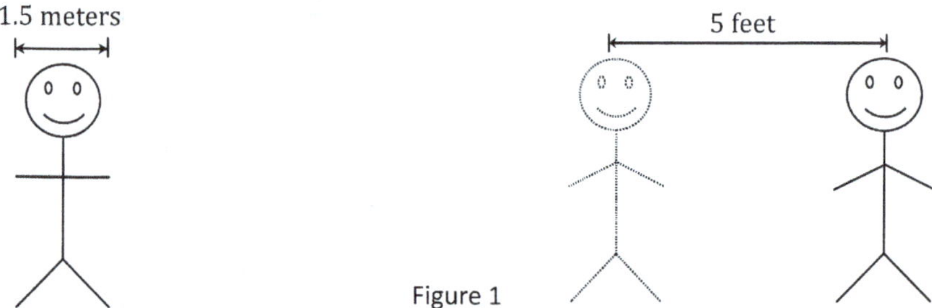

Figure 1

Consider a line segment of length 'l' with one end 'A' fixed to a point and the other end 'B' free to rotate (see figure 2). The curve of length 's' formed by rotating B to B' is called the 'arc' of rotation (see figure 3).

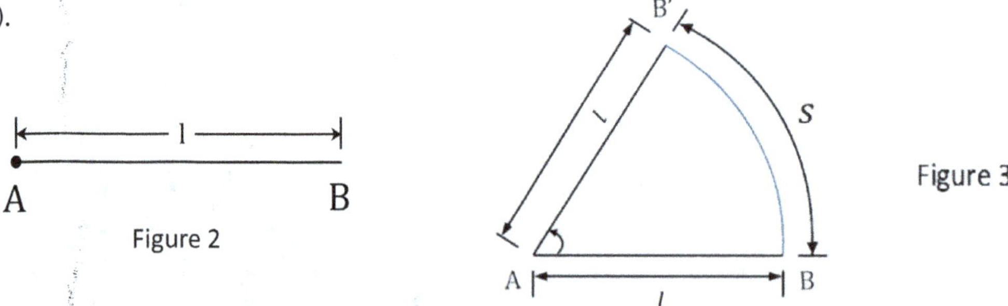

The ratio of the length of the arc (s), to the length of the rotated line (l), is called the 'angle of rotation' or simply 'angle' (θ).

$$\text{angle} = \frac{\text{length of the arc}}{\text{length of the rotated line}} \quad \text{or} \quad \theta = \frac{s}{l} \qquad \text{--- (1)}$$

Since the quantities in the ratio are of the same dimensions, angles by definition are dimensionless. But they are measured in many different units. The most natural one is the SI unit 'radian'(rad). **One radian is obtained when the arc of rotation equals the length of the rotated line.** That is, when $s = l$ (see figure 4).

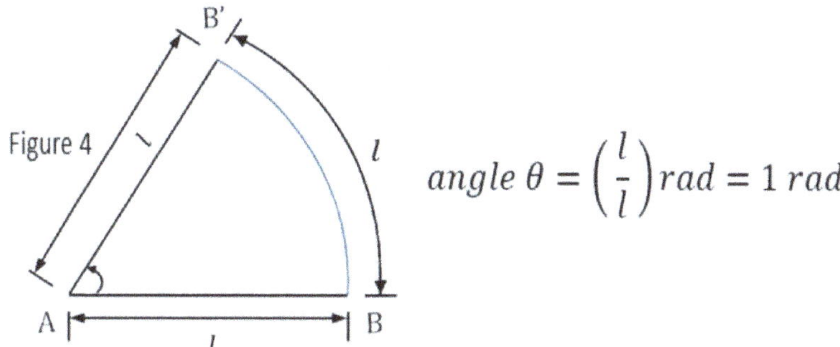

Figure 4

$$angle\ \theta = \left(\frac{l}{l}\right) rad = 1\ rad$$

If we continue rotating, so that the line segment returns to its initial position, we get a circle (see figure 5). This is often called a 'cycle of rotation'. For a complete cycle, the arc is the circumference (c), and the rotated line is the radius (r) (see figure 6).

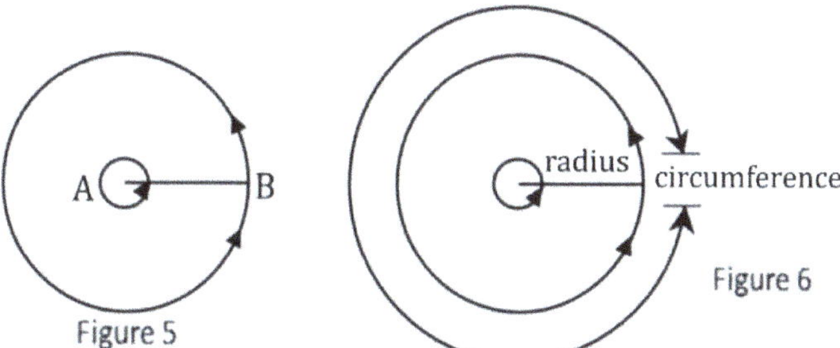

Therefore,

$$\text{angle of a complete cycle} = \frac{\text{circumference}}{\text{radius}} = \left(\frac{c}{r}\right) rad \qquad \text{--- (2)}$$

For every circle, the ratio of the circumference (c) to the diameter (d) is a constant called 'pi', denoted as 'π' and has a value 3.141592… (see figure 7).

i.e. $\dfrac{c}{d} = \pi$ or $\dfrac{c}{2r} = \pi$

or, $c = 2\pi r$ --- (3)

Thus, the angle of a complete cycle becomes,

$$\left(\frac{c}{r}\right) rad = \left(\frac{2\pi r}{r}\right) rad = 2\pi\ rad \qquad \text{--- (4)}$$

It is natural to extend this method to find other angles. For instance,

$$\text{The angle of a half cycle} = \frac{\text{arc of the half cycle}}{\text{radius}}$$

$$= \left(\frac{\frac{c}{2}}{r}\right) rad = \left(\frac{\frac{2\pi r}{2}}{r}\right) rad = \pi\ rad \text{ (see figure 8).} \quad \text{--- (5)}$$

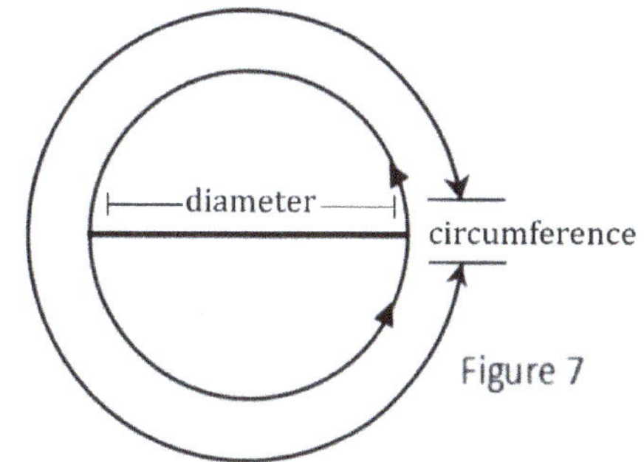

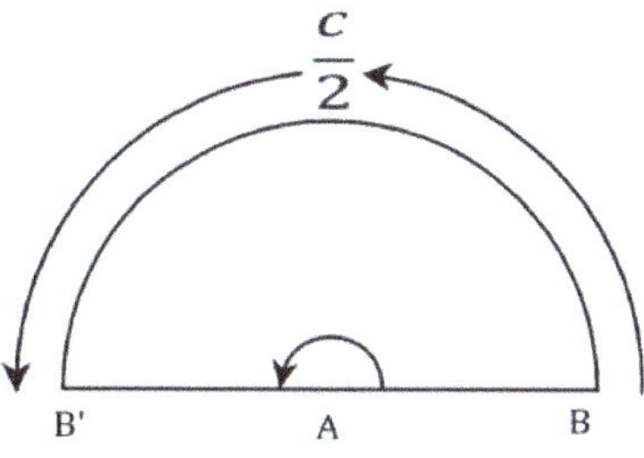

Angle of a half cycle
Figure 8

> <u>**Problem 1**</u>: Draw the following rotations and find the angles corresponding to them.
>
>   i. $\frac{1}{6}$ th of a cycle   ii. $\frac{1}{4}$ th of a cycle   iii. $\frac{3}{4}$ th of a cycle
>
>   iv.   **2 cycles**   v.   **0 cycles**
>
> <u>**Ans**</u>:
>
>   i. $\frac{\pi}{3}$ rad   ii. $\frac{\pi}{2}$ rad   iii. $\frac{3\pi}{2}$ rad
>
>   iv.   **4π rad**   v.   **0 rad**†
>
> † For no rotation, arc = 0, therefore angle = 0 rad

NOTE:

If $\theta = \frac{\pi}{2}$; the angle is called right.

If $\theta < \frac{\pi}{2}$; the angle is called acute.

If $\theta > \frac{\pi}{2}$; the angle is called obtuse.

If $\theta > \pi$; the angle is called oblique (se

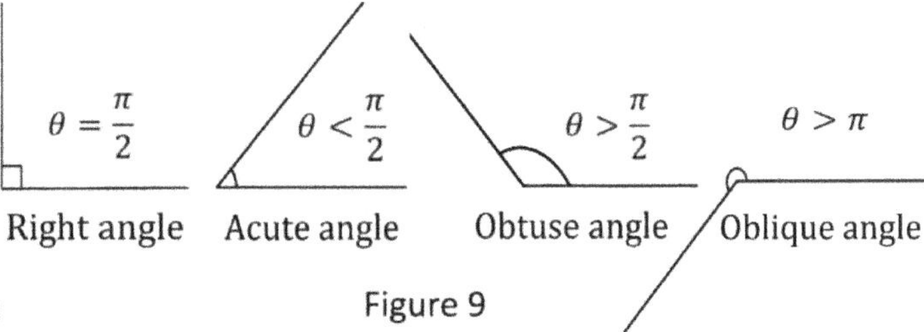

Figure 9

RADIAN V/S DEGREE

Even though the SI unit of angle is radian, a number of other units are also used in practice. The most common unit is called **'degree'** (°). The number of **degree units equivalent to one complete cycle is 360** (see figure 10).

Okay, but why 360, why not 361, or 153, or any other number of choices. Why 360 in particular.

One assumption is that the ancient astronomers approximated a year as 360 days and believed that, completing a year is similar to completing a cycle.

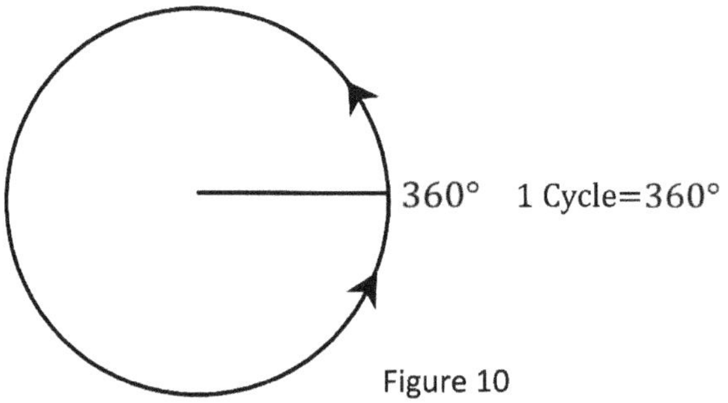

Figure 10

Another assumption is that the Babylonians used a number system with Base 60 (sexagesimal), and 360 is a multiple of 60 with a large number of factors. So, if a complete rotation is 360°, it is easy to divide it into sectors (see figure 11).

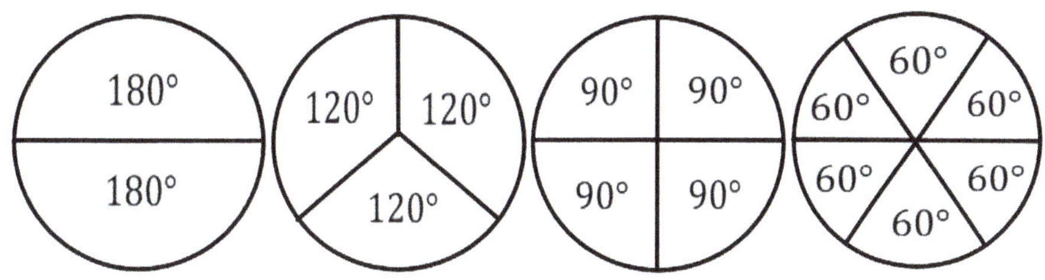

Figure 11

Anyways, long story short, it is the way it is.

Since, for one complete cycle (from equation (4)),

$$\theta = 2\pi \text{ rad}$$

we have the conversion equation,

$$2\pi \text{ rad} = 360° \qquad \text{--- (6)}$$

$$\text{or,} \qquad 1 \text{ rad} = \frac{180°}{\pi} \qquad \text{--- (7)}$$

$$\text{also,} \qquad 1° = \frac{\pi}{180} \text{ rad} \qquad \text{--- (8)}$$

If you are not familiar with conversion equations, see the box below. If you are, then skip it.

HOW TO WORK WITH CONVERSTION EQUATIONS

Conversion equations connect the same quantity, expressed in two different units.

For instance, 2π rad and $360°$ represents the same angle in two different units.

$$2\pi \text{ rad} = 360° \qquad \text{--- (6)}$$

This equation can be manipulated to find other angles in the same units. For that, we should write the equation in a different manner. We know that,

$$2\pi \text{ times 1 radian} = 2\pi \text{ radian}$$

$$\text{and} \qquad 360 \text{ times 1 degree} = 360 \text{ degree}$$

Therefore, the equation (6) becomes,

$$2\pi(1 \text{ rad}) = 360(1°)$$

From this form, we can find the degree units corresponding to 1 radian, just divide by 2π.

$$\text{i.e} \quad 1 \text{ rad} = \frac{360}{2\pi}(1°) = \frac{180°}{\pi} \qquad \text{--- (7)}$$

We can also find the radian units corresponding to 1 degree, just divide by 360.

$$1° = \frac{2\pi}{360}(1\text{rad}) = \frac{\pi}{180}\text{rad} \quad \text{--- (8)}$$

To summarize, we have,

$$\boxed{\begin{aligned} 1\text{rad} &= \frac{180°}{\pi} \\ 1° &= \frac{\pi}{180}\text{rad} \end{aligned}}$$

From these two equations, one can find the degree and radian measures corresponding to any angle of choice. For instance, to find the degrees corresponding to $\frac{\pi}{2}$ radian, just multiply the first equation by $\frac{\pi}{2}$.

$$\text{i.e} \quad \frac{\pi}{2}(1\text{ rad}) = \frac{\pi}{2}\left(\frac{180°}{\pi}\right)$$

$$\text{or,} \quad \frac{\pi}{2}\text{rad} = 90°$$

Or, to find the radians corresponding to 180°, just multiply the second equation by 180,

$$\text{i.e} \quad 180(1°) = 180\left(\frac{\pi}{180}\right)\text{rad}$$

$$\text{or,} \quad 180° = \pi\text{ rad}$$

This method can be easily generalized and applied to any unit conversions.

Problem: If 2 US Dollars (\$) = 212.32 Japanese Yen (¥). Find out how much dollars will be equivalent to 15 Yen.

Ans: .141296

<u>Problem 2</u>: Find the degree measures corresponding to the following radians. Also draw the corresponding rotations.

   i. $\frac{\pi}{3}$ rad   ii. $\frac{\pi}{4}$ rad   iii. $\frac{\pi}{6}$ rad   iv. $\frac{3\pi}{2}$ rad

   v. $\frac{2\pi}{3}$ rad   vi. 2π rad   vii. 4π rad

<u>Ans</u>:

   i.   60°   ii.   45°   iii.   30°   iv.   270°   v.   120°

   vi.   360°   vii.   720°

Now, at this point, it is good to take a pause and reflect on your progress. If you are satisfied with the topics and was able to do the problems, take a break, eat something if you are hungry, meditate for a while, and continue. If you are not satisfied, take a break, meditate, and find out the exact difficulty. Read, think, and read again until you understand it. Remember, continue only if you are clear about the topics. Forget the "one-week" thing, it's a publicity stunt anyway.

SESSION TWO

<u>CHOPPING UP A DEGREE</u>

To measure lengths that are shorter than a meter, we use the unit 'centimeter' (cm), which is the one-hundredth fraction of one meter. It is a more practical unit for many day-to-day measurements. Similarly, to measure angles that are smaller than one degree, we use the unit **'minute' (')**, which is the **one-sixtieth fraction of one degree** (see figure 12).

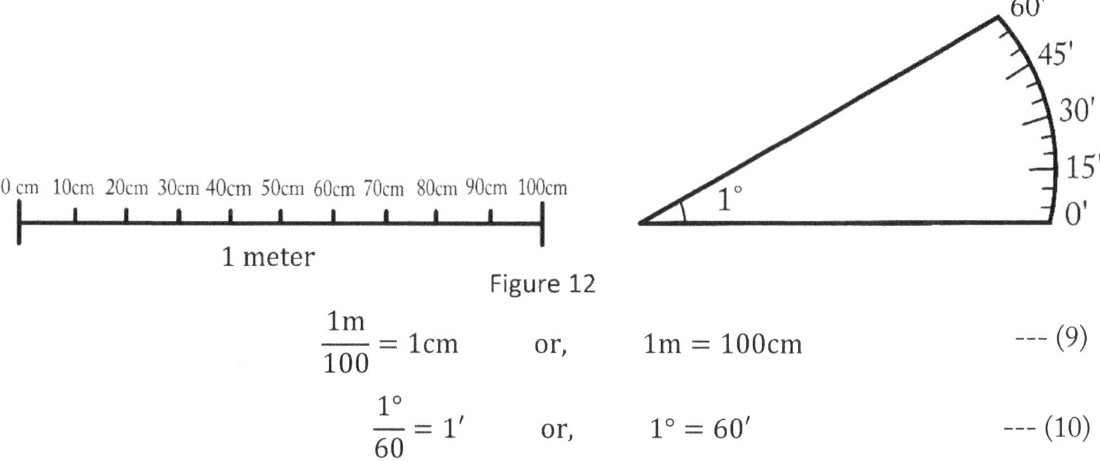

Figure 12

$$\frac{1m}{100} = 1cm \qquad or, \qquad 1m = 100cm \qquad \text{--- (9)}$$

$$\frac{1°}{60} = 1' \qquad or, \qquad 1° = 60' \qquad \text{--- (10)}$$

These units increase the precision of our calculations by enabling us to measure lengths and angles that are smaller than a meter and a degree. For instance,

$$\text{half of a metre} = \frac{1}{2}(1m) = \frac{1}{2}(100cm) = 50cm$$

$$or, \quad \text{quarter of a metre} = \frac{1}{4}(1m) = 25cm$$

Okay, how about a length of 2 meters and 45 centimeters (2m45cm). Here, we have to add 45 centimeters to 2 meters. That is,

$$2m45cm = 2m + 45cm$$

To add terms of different units, we should convert one of them to the other. That is,

$$2\text{m}45\text{cm} = 2\text{m} + 45\text{cm}$$

$$= 2\text{m} + \left(\frac{45}{100}\right)\text{m}$$

$$= 2\text{m} + .45\text{m}$$

$$= 2.45\text{m} \quad [\text{Note: the answer is in meters}]$$

$$\text{Also,} \quad 2\text{m}45\text{cm} = 2\text{m} + 45\text{cm}$$

$$= 200\text{cm} + 45\text{cm}$$

$$= 245\text{cm} \quad [\text{Note: the answer is in centimeters}]$$

> **Key point: To add two different units, we have to make them same**

The same principle can be applied to angles. For instance,

$$2°45' = 2° + 45'$$

$$= 2° + \frac{45°}{60}$$

$$= \frac{33°}{12} \quad [\text{Note: the answer is in degrees}]$$

$$\text{Also,} \quad 2°45' = 2° + 45'$$

$$= 120' + 45'$$

$$= 165' \quad [\text{Note: the answer is in minutes}]$$

Example 1: Convert 2°45' into radians.

Solution:

$$2°45' = 2° + \frac{45°}{60} = \frac{33°}{12}$$

The degree to radian conversion is given by

$$1° = \frac{\pi}{180}\,\text{rad}$$

Therefore,

$$\frac{33°}{12} = \left(\frac{33}{12}\right)\frac{\pi}{180}\,\text{rad} = \frac{11\pi}{720}\,\text{rad}$$

Problem 3: Find the radian measures corresponding to the following degrees.

    i. $30°$ ii. $40°20'$ iii. $300°$ iv. $(-37°30')^{\dagger}$

    v. $110°30'$

Ans:

    i. $\dfrac{\pi}{6}\,\text{rad}$ ii. $\dfrac{121\pi}{540}\,\text{rad}$ iii. $\dfrac{5\pi}{3}\,\text{rad}$ iv. $-\dfrac{5\pi}{24}\,\text{rad}$

    v. $\dfrac{221\pi}{360}\,\text{rad}$

$\dagger$ Negative sign means the rotation is in the clockwise direction.

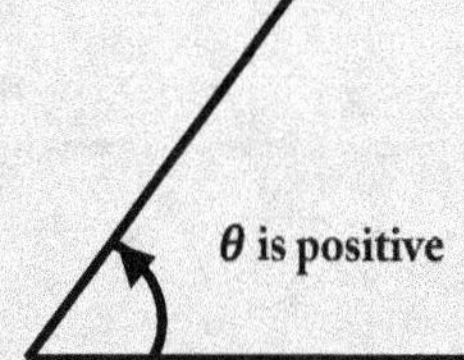

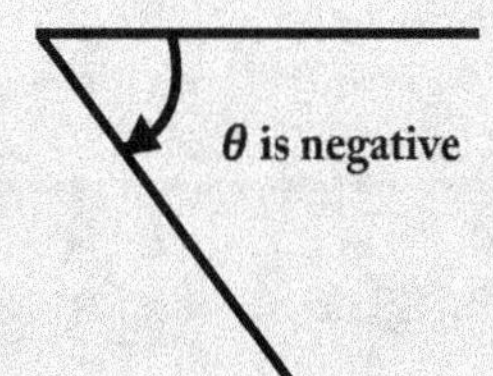

COUNTING SECONDS

In the previous session, we have defined angles, expressed them in radians and degrees. In this session, we have divided a degree into sixty slices called minutes.

Now, we go even further by dividing a minute into smaller slices of angles called '**seconds**' ("). **A second is the one-sixtieth fraction of a minute.**

$$\frac{1'}{60} = 1'' \qquad \text{or,} \qquad 1' = 60'' \qquad\qquad \text{--- (11)}$$

From equation (10) we have,

$$1' = \frac{1°}{60} \qquad\qquad \text{--- (10)}$$

Combining with equation (11) we get,

$$\frac{1°}{60} = 60''$$

Thus, we have the **second to degree** conversion equation,

$$1'' = \frac{1°}{3600} \qquad\qquad \text{--- (12)}$$

Example 2: Convert 25°15'30" into radians.

Solution:

$$25°15'30'' = 25° + \frac{15°}{60} + \frac{30°}{3600} = \frac{3031°}{120}$$

Using the degree to radian conversion equation,

$$1° = \frac{\pi}{180}\,\text{rad}$$

We get,

$$\frac{3031°}{120} = \frac{3031\pi}{21600}\,\text{rad}$$

Problem 4: Express the following angles in radians.

    i.   47°25'36" ii.   52°30'24" iii.   64°24'38" iv.   228°23'10"

Ans:

    i. $\dfrac{3557\pi}{13500}\,\text{rad}$ ii. $\dfrac{1969\pi}{6750}\,\text{rad}$ iii. $0.35783\pi\,\text{rad}$

    iv. $1.2688\pi\,\text{rad}$

Now we know how to convert angles written in terms of degrees, minutes, and seconds into radian.

But the reverse is somewhat trickier. Let's look at an example.

Example 3: Convert 8 radians into degrees, minutes and seconds.

Solution: The radian to degree conversion is,

$$1\,\text{rad} = \frac{180°}{\pi}$$

Therefore,

$$8\,\text{rad} = 8\left(\frac{180°}{\pi}\right) = 458.3662° \quad [\text{Where } \pi \approx 3.141592]$$

Now split the angle into decimal and non-decimal part. That is,

$$458.3662° = 458° + .3662°$$

Convert the decimal part into minutes.

$$.3662° = .3662 \times 60' = 21.972'$$

Therefore,

$$458° + .3662° = 458° + 21.972'$$

Again, split the decimal and non-decimal part.

$$458° + 21' + .972'$$

Convert the decimal part into seconds.

$$.972' = .972 \times 60'' = 58.32''$$

Thus, we get,

$$458° + 21' + .972' = 458° + 21' + 58.32'' = 458°21'58.32''$$

Problem 5: **Convert the following radians into degrees, minutes and seconds.**

i. 1 rad ii. 3 rad iii. 5 rad iv. $\dfrac{1969\pi}{6750}$ rad

Ans:

i. 57°17'44.8" ii. 171°53'14.419" iii. 286°28'44.03" iv. 52°30'24"

That's it, we are done with angles and their conversions. Now, take a few minutes off. Drink some water, have some nuts, meditate, and recall everything from sessions one and two. If you are confident about the topics, continue studying. If you are not, then meditate, clear your mind, drink some water if you want, and return to the same session.

SESSION THREE

WHAT IS TRIGONOMETRY

We studied lengths, we studied angles. Now it's time to combine them both. **Structures having both lengths and angles are called polygons.** A square is a polygon with four sides of equal length and four angles of equal magnitude (see figure 13).

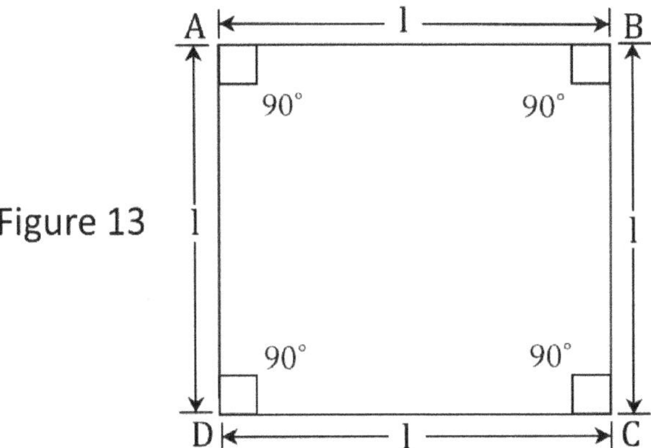

The four angles are formed at the four vertices A, B, C and D, and each of them have a magnitude of 90 degrees.

A triangle is a polygon having only three sides. The sides are not necessarily equal. If they are equal, then it's an equilateral (see figure 14 and 15).

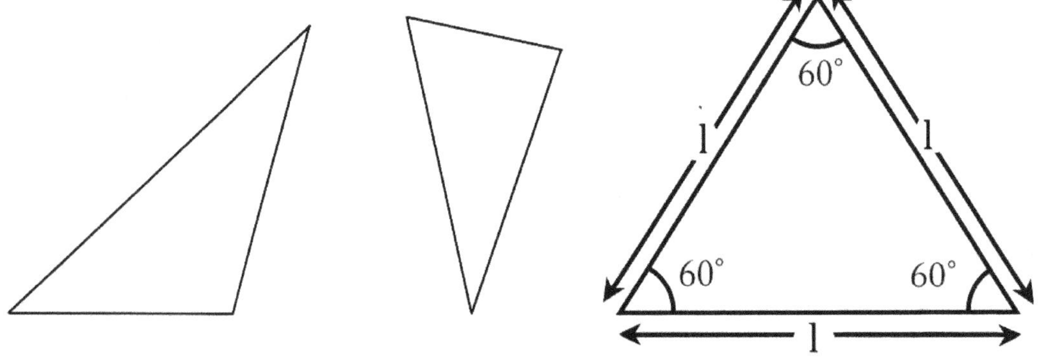

Figure 14   Figure 15

For an equilateral triangle, the three angles are also equal. Otherwise, they are not. The most fundamental theorem concerning a triangle is that **the sum of all angles equals 180 degrees**. That makes the angles of an equilateral triangle 60 degrees each (see figure15).

A triangle with one of the angles 90° is called a right-angled triangle, or simply right triangle.

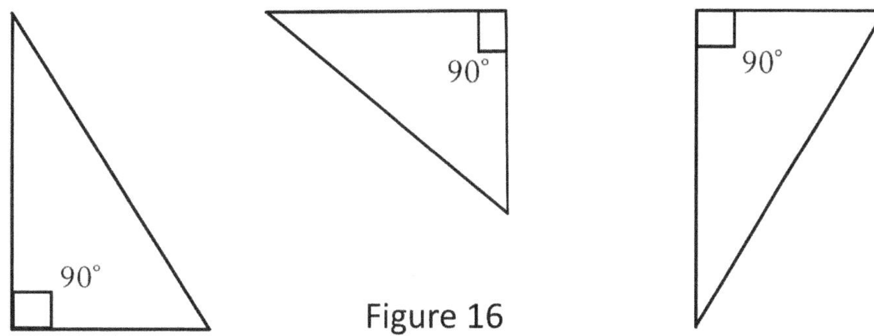

Figure 16

Historically, trigonometry was conceived as the study of triangles, in particular, right triangles. But the invention of coordinate geometry extended its scope and applications into much larger domains. How it does that "is one or the many mysteries that this book will reveal".

But for now, let's grapple with some basic properties of right triangles so that the transition to coordinate trigonometry becomes effortless.

Consider a right-angled triangle having three angles at the three vertices A, B and C (see figure 17). The angle at the vertex B is the right angle. The side opposite to the right angle is called the **'hypotenuse'.** It is the largest side of a right-angled triangle. For triangle ABC it is the side AC (see figure 18).

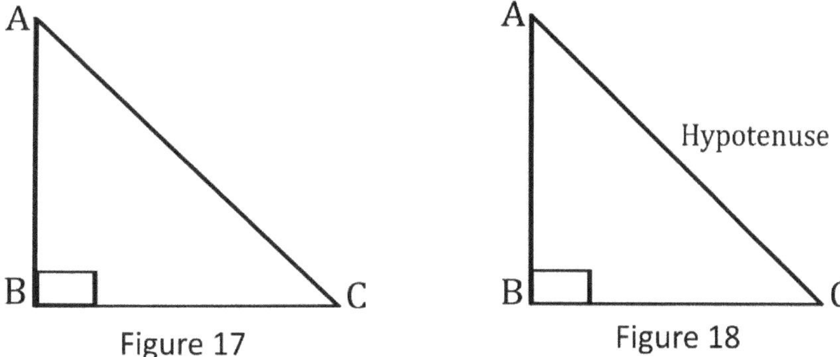

Figure 17   Figure 18

It is interesting to note that the entire subject of trigonometry springs out of a single fundamental theorem called the **Pythagoras theorem.** It states that **the square of the hypotenuse equals the sum of squares of the other two sides.** That is,

$$(AC)^2 = (AB)^2 + (BC)^2 \qquad\qquad \text{--- (13)}$$

Therefore, the hypotenuse is,

$$AC = \sqrt{AB^2 + BC^2} \qquad\qquad \text{--- (14)}$$

$$\text{Also} \quad AB = \sqrt{AC^2 - BC^2} \qquad\qquad \text{--- (15)}$$

$$\text{And} \quad BC = \sqrt{AC^2 - AB^2} \qquad\qquad \text{--- (16)}$$

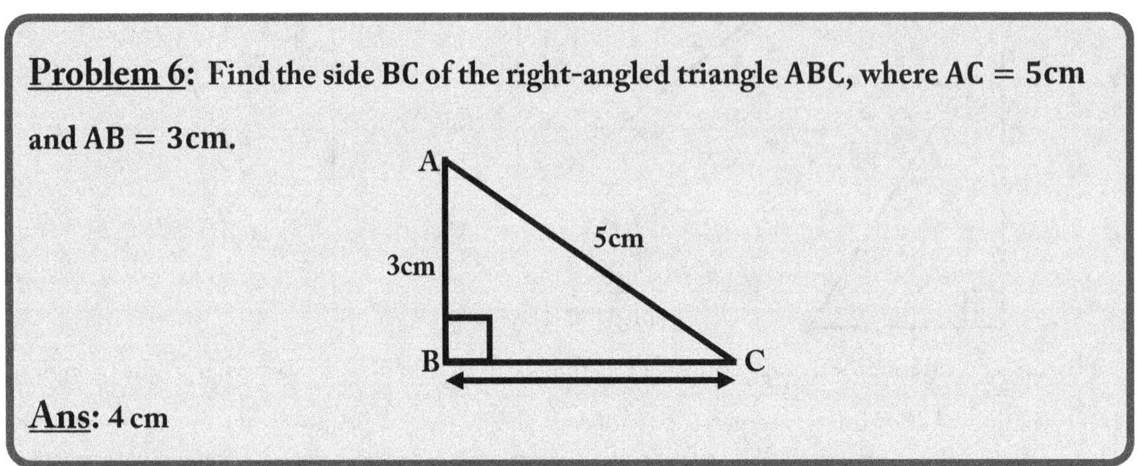

<u>**Problem 6**</u>: **Find the side BC of the right-angled triangle ABC, where AC = 5cm and AB = 3cm.**

<u>**Ans**</u>: 4 cm

TRIGONOMETRIC FUNCTIONS

Again, consider the right-angled triangle ABC. Let the angle at the vertices A and C be named 'θ' and 'φ' respectively (see figure 19).

Figure 19

BC is the side opposite to the angle θ and AB is the side adjacent to it. Then the basic trigonometric functions **Sine, Cosine, Tangent** and their respective reciprocals **Cosecant, Secant and Cotangent** are defined as,

$$\text{Sine of } \theta = \text{Sin}(\theta) = \frac{\text{opposite side of } \theta}{\text{hypotenuse}} = \frac{\text{BC}}{\text{AC}} \qquad \text{--- (17)}$$

$$\text{Cosine of } \theta = \text{Cos}(\theta) = \frac{\text{adjacent side of } \theta}{\text{hypotenuse}} = \frac{\text{AB}}{\text{AC}} \qquad \text{--- (18)}$$

$$\text{Tangent of } \theta = \text{Tan}(\theta) = \frac{\sin(\theta)}{\cos(\theta)} = \frac{\text{opposite side of } \theta}{\text{adjacent side of } \theta} = \frac{\text{BC}}{\text{AB}} \qquad \text{--- (19)}$$

$$\text{Cosecant of } \theta = \text{Cosec}(\theta) = \frac{1}{\sin(\theta)} = \frac{\text{hypotenuse}}{\text{opposite side of } \theta} = \frac{\text{AC}}{\text{BC}} \qquad \text{--- (20)}$$

$$\text{Secant of } \theta = \text{Sec}(\theta) = \frac{1}{\cos(\theta)} = \frac{\text{hypotenuse}}{\text{adjacent side of } \theta} = \frac{\text{AC}}{\text{AB}} \qquad \text{--- (21)}$$

$$\text{Cotangent of } \theta = \text{Cot}(\theta) = \frac{1}{\tan(\theta)} = \frac{\cos(\theta)}{\sin(\theta)} = \frac{\text{adjacent side of } \theta}{\text{opposite side of } \theta} = \frac{\text{AB}}{\text{BC}} \text{--- (22)}$$

Looks like a lot to take in, right. But, actually it's pretty easy. Try to remember at least one equation, and the rest can be easily deduced.

Look at the first equation,

$$\text{Sin}(\theta) = \text{Opposite over Hypotenuse}$$

To memorise this equation, remember the phrase:

Standing On the Hill

or,

Signal On the Highway

or,

Any phrase with the letters **S, O, H.**

If Sine is committed to memory, deducing Cosine is easy, just replace 'Opposite' by 'Adjacent'. If Sine is difficult to memorise, look at the second equation,

$$\text{Cos}(\theta) = \text{Adjacent over Hypotenuse}$$

To memorise this equation, remember the phrase:

Camel And Hippo

or,

Coffee And Hotdog (or Hotcake)

or,

Any phrase with the letters **C, A, H.**

To find Sine from Cosine, replace 'Adjacent' by 'Opposite'. You have to remember at least one of the two equations, there is no way getting around it. The best way to memorise an equation is by the 'flash card' technique described in the preface. Vocal repetition also helps boost your memory. So just sing it casually. If you have $\text{Sin}(\theta)$ and $\text{Cos}(\theta)$ in your pocket, the rest can be easily deduced. $\text{Tan}(\theta)$ is just the ratio of $\text{Sin}(\theta)$ and $\text{Cos}(\theta)$. $\text{Cosec}(\theta)$, $\text{Sec}(\theta)$ and $\text{Cot}(\theta)$ are the reciprocals of $\text{Sin}(\theta)$, $\text{Cos}(\theta)$ and $\text{Tan}(\theta)$ respectively.

Problem 7: Write the basic trigonometric functions of the angle 'φ' in the above triangle ABC (see figure 19) in terms of its sides.

Ans:

$$\sin \varphi = \frac{AB}{AC} \qquad \cos \varphi = \frac{BC}{AC} \qquad \tan \varphi = \frac{AB}{BC}$$

$$\operatorname{cosec} \varphi = \frac{AC}{AB} \qquad \sec \varphi = \frac{AC}{BC} \qquad \cot \varphi = \frac{BC}{AB}$$

Problem 8: Write the basic trigonometric functions of the angle 'θ' in the following triangles.

i. 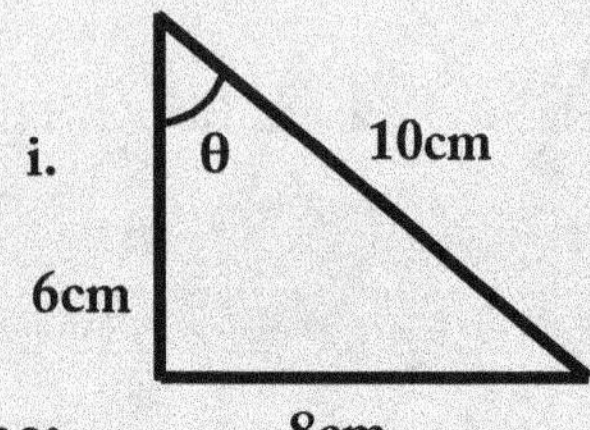

ii. 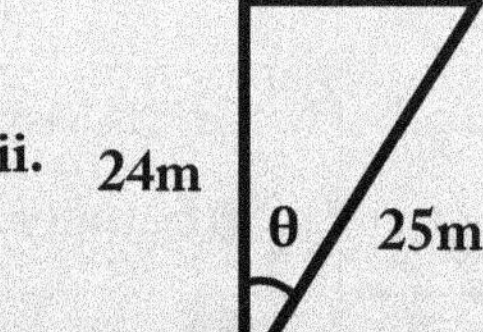

iii. 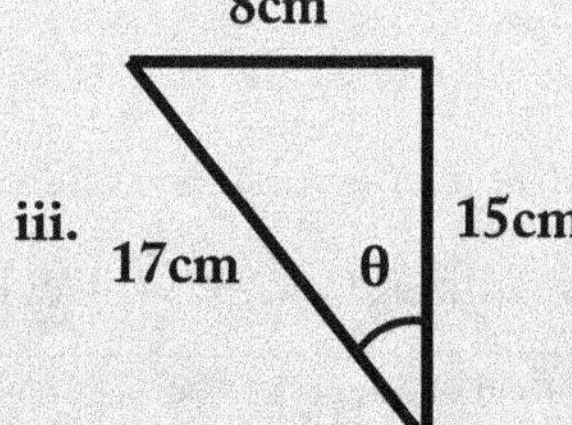

Ans:

i. $\sin \theta = \dfrac{8}{10} \quad \cos \theta = \dfrac{6}{10} \quad \tan \theta = \dfrac{8}{6} \quad \operatorname{cosec} \theta = \dfrac{10}{8} \quad \sec \theta = \dfrac{10}{6} \quad \cot \theta = \dfrac{6}{8}$

ii. $\sin \theta = \dfrac{7}{25} \quad \cos \theta = \dfrac{24}{25} \quad \tan \theta = \dfrac{7}{24} \quad \operatorname{cosec} \theta = \dfrac{25}{7} \quad \sec \theta = \dfrac{25}{24} \quad \cot \theta = \dfrac{24}{7}$

iii. $\sin \theta = \dfrac{8}{17} \quad \cos \theta = \dfrac{15}{17} \quad \tan \theta = \dfrac{8}{15} \quad \operatorname{cosec} \theta = \dfrac{17}{8} \quad \sec \theta = \dfrac{17}{15} \quad \cot \theta = \dfrac{15}{8}$

Having studied lengths, angles, the Pythagoras theorem and the trigonometric functions, we have acquired the basic toolkit to start doing trigonometry. But for today let's wrap it up.

Before ending the session, remember to meditate and recall everything. If you are satisfied with the sessions, you are good to go. Have a good night sleep and comeback tomorrow. If you are not satisfied, its ok, tomorrow return to the same session.

Don't give up

Day Two

Session One
Basic identities and
how to memorize them

Session Two
More identities and how to prove them

Session Three
More proofs

I hate every minute of my training. But
I said, don't quit. Suffer now and live
the rest of your life as a champion.

- Muhammad Ali

SESSION ONE

Take a few breaths and meditate on everything we have studied so far. If you have flash cards, go through them with patience and focus. Try to recall and reconnect with each and every topic.

Today, we will learn the basics trigonometric identities, and how to use them to prove other identities. Memorising identities is as important as learning them. So, try to memorise them thoroughly. If you are having difficulties in memorising, try to learn the shortcut presented at the end.

BASIC IDENTITIES AND HOW TO MEMORIZE THEM

We first present the proof of the identities. If you are not interested in the proof, skip it and learn the equations (25), (26) and (27).

Let's begin,

Consider the right-angled triangle ABC, with the angle θ at the vertex A.

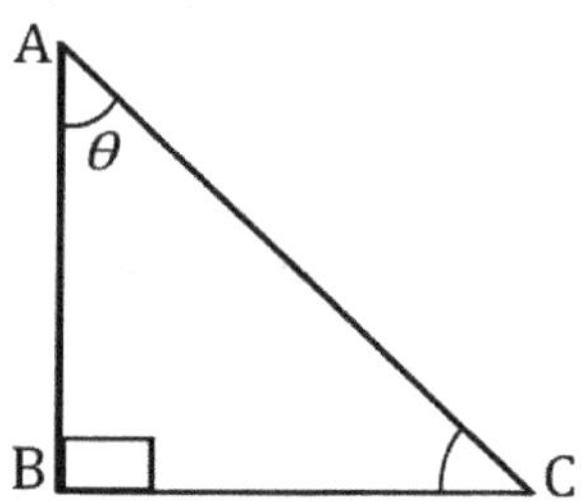

$$\sin(\theta) = \frac{BC}{AC} \qquad \text{Therefore,} \qquad BC = (AC)\sin(\theta) \qquad \text{--- (23)}$$

$$\cos(\theta) = \frac{AB}{AC} \qquad \text{Therefore,} \qquad AB = (AC)\cos(\theta) \qquad \text{--- (24)}$$

Substituting into the Pythagoras theorem,

$$(AC)^2 = (BC)^2 + (AB)^2 \qquad \text{--- (13)}$$

We get,

$$(AC)^2 = (AC)^2\left(\sin(\theta)\right)^2 + (AC)^2\left(\cos(\theta)\right)^2$$

It is customary to write,

$$\left(\sin(\theta)\right)^2 \text{ as } \sin^2\theta \text{ and } \left(\cos(\theta)\right)^2 \text{ as } \cos^2\theta$$

[It applies to other functions also, e.g. $\left(\tan(\theta)\right)^2 = \tan^2\theta$ and so on.]

Therefore, $\quad (AC)^2 = (AC)^2 \sin^2\theta + (AC)^2 \cos^2\theta$

$\Rightarrow \quad (AC)^2 = (AC)^2(\sin^2\theta + \cos^2\theta)$

or, $\qquad \boxed{\sin^2\theta + \cos^2\theta = 1}$ --- (25)

This is the first trigonometric identity.

Dividing the equation (25) by $\cos^2\theta$, we get,

$$\frac{\cos^2\theta + \sin^2\theta}{\cos^2\theta} = \frac{1}{\cos^2\theta}$$

$$\Rightarrow \quad 1 + \frac{\sin^2\theta}{\cos^2\theta} = \frac{1}{\cos^2\theta}$$

Since,

$$\frac{\sin\theta}{\cos\theta} = \tan\theta \quad \text{and} \quad \frac{1}{\cos\theta} = \sec\theta$$

We get the second identity,

$$\boxed{1 + \tan^2\theta = \sec^2\theta}$$ --- (26)

Now, dividing the equation (25) by $\sin^2\theta$, we get,

$$\frac{\sin^2\theta + \cos^2\theta}{\sin^2\theta} = \frac{1}{\sin^2\theta}$$

$$\Rightarrow \quad 1 + \frac{\cos^2\theta}{\sin^2\theta} = \frac{1}{\sin^2\theta}$$

Since,

$$\frac{\cos\theta}{\sin\theta} = \cot\theta \quad \text{and} \quad \frac{1}{\sin\theta} = \operatorname{cosec}\theta$$

We get the third identity,

$$\boxed{1 + \cot^2\theta = \operatorname{cosec}^2\theta}$$ --- (27)

To summarise, we have,

$$\boxed{\begin{aligned} &\mathbf{\sin^2\theta + \cos^2\theta = 1} \\ &\mathbf{1 + \tan^2\theta = \sec^2\theta} \\ &\mathbf{1 + \cot^2\theta = \operatorname{cosec}^2\theta} \end{aligned}}$$

Take your time to memorise them in their original form. No tricks, no shortcuts, and no magic phrases. Just the way they are written above. If you are making flash cards, just write "BASIC IDENTITIES", and try to recall the equations.

But, if you are having a hard time remembering, there is one method you can follow. Keep in mind, this should be used only as an emergency back-up plan.

The idea is to draw a hexagon, which is a polygon with six sides and six vertices (see figure 20). Write $\sin^2 \theta$, $\cos^2 \theta$ and $\tan^2 \theta$ on the vertices as shown below, with the number '1' at the centre.

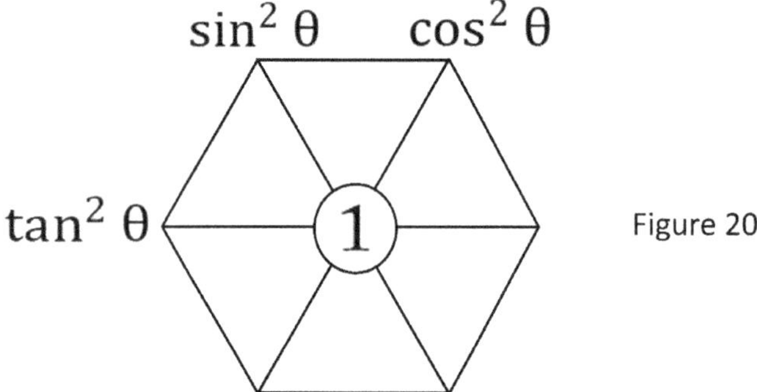

Figure 20

Write their reciprocals at the vertices diametrically opposite to each function (see figure 21).

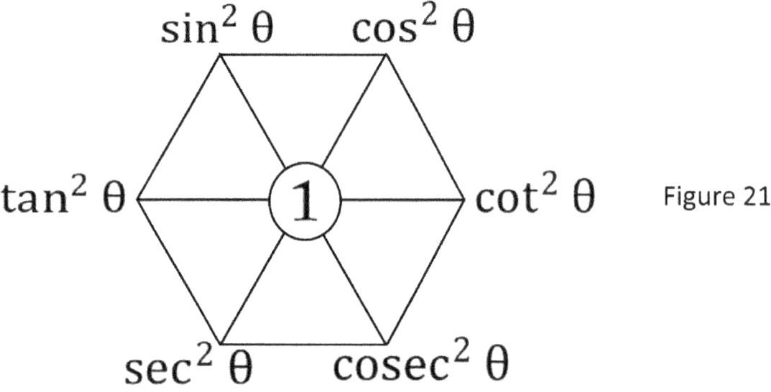

Figure 21

Draw the arrow sign on the hexagon as shown below (see figure 22).

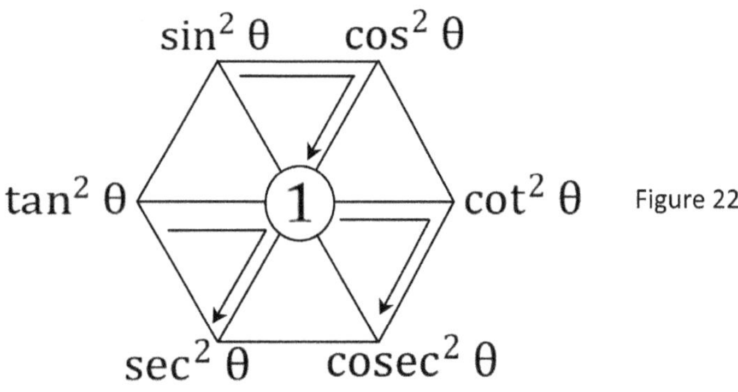

Figure 22

Adding the functions on the tail and the vertex of the arrow gives the LHS. And the function pointed by the arrowhead gives the RHS (see figure 23).

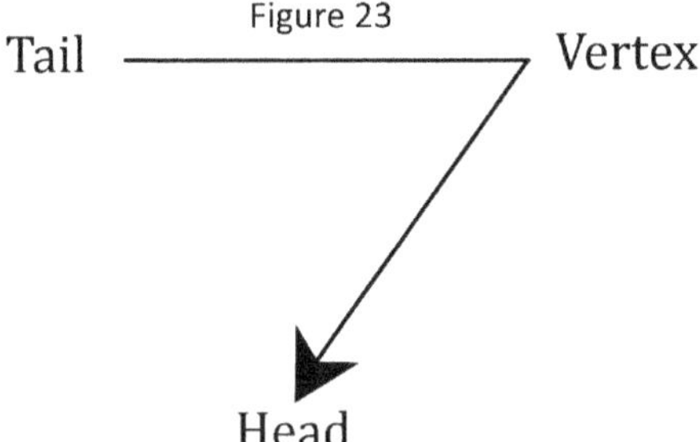

For instance, consider the triangle,

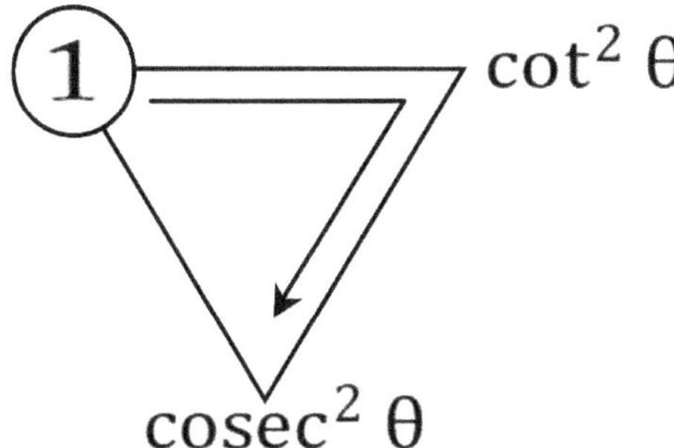

Here, '1' is at the tail of the arrow and '$\cot^2 \theta$' is at the vertex.

Therefore,

$$\mathrm{LHS} = 1 + \cot^2 \theta$$

The arrowhead is pointed towards $\mathrm{cosec}^2 \theta$,

Therefore,

$$\mathrm{RHS} = \mathrm{cosec}^2 \theta$$

Thus,

$$1 + \cot^2 \theta = \mathrm{cosec}^2 \theta$$

Memorising identities is a crucial part of the subject. So, take your time to do so, and continue when you are confident. If you can't, at least have this emergency back-up.

SESSION TWO

MORE IDENTITIES AND HOW TO PROVE THEM

An identity is an equation that is valid for all values of its variable. For instance,

The equation $\sin^2 \theta + \cos^2 \theta = 1$ is valid for all values of θ. Therefore, it is an identity.

The expansion $(a + b)^2 = a^2 + 2ab + b^2$ is valid for all values of 'a' and 'b'. Therefore, it is also an identity.

But, the equation $2x + 3 = 7$ is not an identity, since it is valid only for $x = 2$.

To prove an identity is to establish the equality between its Left-Hand Side [LHS] and the Right-Hand Side [RHS]. This can be done by converting one of the sides into the other using basic algebraic operations. You can either take the LHS and convert it into the RHS or, take the RHS and convert it into the LHS.

$$\text{LHS} \xrightarrow{\text{Converted to}} \text{RHS}$$

$$\text{or}$$

$$\text{RHS} \xrightarrow{\text{Converted to}} \text{LHS}$$

It is essential to be well equipped with all the concepts and formulae needed. In this case, they are the trigonometric functions, trigonometric identities, and the algebraic identities like,

$$(a + b)^2 = a^2 + 2ab + b^2$$
$$(a - b)^2 = a^2 - 2ab + b^2$$
$$a^2 - b^2 = (a + b)(a - b)$$

Let's look at an example.

Example 4: Prove that $\sin \theta \cot \theta = \cos \theta$.

Solution: Here,

$$\text{LHS} = \sin \theta \cot \theta$$

We know that,

$$\cot \theta = \frac{\cos \theta}{\sin \theta}$$

Therefore, we can write the LHS as,

$$LHS = \sin\theta\cot\theta = \sin\theta\,\frac{\cos\theta}{\sin\theta} = \cos\theta\,, \text{ which is the RHS}$$

We thus took the LHS and converted it into RHS, thereby establishing their equality and completing the proof.

<u>Problem 9</u>: Prove the following identities.

   i. $\cos\theta\cosec\theta = \cot\theta$ ii. $\sec\theta\cot\theta = \cosec\theta$ iii. $\tan\theta\cos\theta = \sin\theta$

   iv. $\sin\theta\sec\theta\cot\theta = 1$ v. $\cos\theta\tan\theta\cosec\theta = 1$

Proving identities often involves algebraic operations and algebraic identities. See the example.

<u>Example 5</u>: Prove that $(1 + \cos\theta)(1 - \cos\theta) = \sin^2\theta.$

<u>Solution</u>: Here,

$$LHS = (1 + \cos\theta)(1 - \cos\theta)$$

We have the algebraic identity,

$$(a + b)(a - b) = a^2 - b^2$$

Therefore, we can write the LHS as,

$$LHS = (1 + \cos\theta)(1 - \cos\theta) = 1^2 - \cos^2\theta$$

Since,
$$1 - \cos^2\theta = \sin^2\theta$$

We have,
$$LHS = 1 - \cos^2\theta = \sin^2\theta = RHS$$

Thus, the proof is complete.

We can also take the RHS and convert it into the LHS, i.e.

$$RHS = \sin^2\theta = 1 - \cos^2\theta = (1 + \cos\theta)(1 - \cos\theta) = LHS$$

Both methods are equally valid.

Before starting a problem, take a few breaths, be calm and stay focused. Try to come up with different ways by which you can make your first move, and write them down. Choose one of them and follow your instincts. If it's not working, choose the next one and continue. If you get exhausted, stop there, drink some water, have some nuts, take a nap if you're tired, and then comeback.

Writing proofs can be tough. You may get stuck, unable to move forward, or you may get lost in some heavy algebra. The idea is to keep practicing. The more you work out, the more you become fluent.

<u>Problem 10</u>: Prove the following identities.

   i. $(\sin\theta + \cos\theta)^2 = 1 + 2\sin\theta\cos\theta$

   ii. $(\sec\theta + 1)(\sec\theta - 1) = \tan^2\theta$

   iii. $\sec\theta\,(\sec\theta + 2\sin\theta) = (\tan\theta + 1)^2$

   iv. $\operatorname{cosec}\theta\,(\operatorname{cosec}\theta - 2\cos\theta) = (\cot\theta - 1)^2$

   v. $(\tan\theta + 1)^2 + (\tan\theta - 1)^2 = 2\sec^2\theta$

   vi. $(\cot\theta - 1)^2 + (\cot\theta + 1)^2 = 2\operatorname{cosec}^2\theta$

   vii. $(\operatorname{cosec}\theta + 1)(\operatorname{cosec}\theta - 1) = \cot^2\theta$

   viii. $\sec\theta - \sec\theta\sin^2\theta = \cos\theta$

SESSION THREE

MORE PROOFS

Let's look at some more examples.

Example 6: Prove that

$$\frac{\sin \theta}{1 + \cos \theta} + \frac{1 + \cos \theta}{\sin \theta} = 2 \operatorname{cosec} \theta.$$

Solution: We have,

$$\text{LHS} = \frac{\sin \theta}{1 + \cos \theta} + \frac{1 + \cos \theta}{\sin \theta}$$

Upon cross multiplication,

$$\text{LHS} = \frac{\sin^2 \theta + (1 + \cos \theta)^2}{\sin \theta \,(1 + \cos \theta)} = \frac{\sin^2 \theta + \cos^2 \theta + 1 + 2 \cos \theta}{\sin \theta \,(1 + \cos \theta)}$$

$$= \frac{1 + 1 + 2 \cos \theta}{\sin \theta \,(1 + \cos \theta)} = \frac{2 + 2 \cos \theta}{\sin \theta \,(1 + \cos \theta)}$$

$$= \frac{2(1 + \cos \theta)}{\sin \theta \,(1 + \cos \theta)} = \frac{2}{\sin \theta} = 2 \operatorname{cosec} \theta = \text{RHS}.$$

Problem 11: Prove the following identities.

i. $\dfrac{\sin \theta}{1 - \cos \theta} + \dfrac{1 - \cos \theta}{\sin \theta} = 2 \operatorname{cosec} \theta$

ii. $\dfrac{1}{1 - \sin A} + \dfrac{1}{1 + \sin A} = 2 \sec^2 A$

iii. $\cot \theta + \dfrac{\sin \theta}{1 + \cos \theta} = \operatorname{cosec} \theta$

iv. $\dfrac{\operatorname{cosec} \theta}{\operatorname{cosec} \theta - 1} + \dfrac{\operatorname{cosec} \theta}{\operatorname{cosec} \theta + 1} = 2 \sec^2 \theta$

v. $\dfrac{1 + \sin x}{\cos x} + \dfrac{\cos x}{1 + \sin x} = 2 \sec x$

vi. $\dfrac{\sec \theta}{\sec \theta + 1} + \dfrac{\sec \theta}{\sec \theta - 1} = 2 \operatorname{cosec}^2 \theta$

vii. $(\operatorname{cosec} \theta - \sin \theta)(\sec \theta - \cos \theta)(\tan \theta + \cot \theta) = 1$

viii. $(1 - \cos\theta)(1 + \sec\theta)\cot\theta = \sin\theta$ **ix.** $\dfrac{\sin\theta}{\csc\theta} + \dfrac{\cos\theta}{\sec\theta} = 1$

x. $\sec^2 x + \csc^2 x = \sec^2 x \csc^2 x$

Before ending the session, we shall go through one last method called the '**conjugate multiplication**'. The conjugate of a two-term expression is obtained when the sign between the terms is reversed. For instance, the conjugate of $a + b$ is $a - b$ and the conjugate of $a - b$ is $a + b$. See the example.

<u>Example 7</u>: Prove that

$$\frac{\cos\theta}{1 + \sin\theta} = \frac{1 - \sin\theta}{\cos\theta}$$

<u>Solution</u>: We have,

$$LHS = \frac{\cos\theta}{1 + \sin\theta}$$

Here the conjugate of the denominator is $1 - \sin\theta$.

The method is to multiply both numerator and denominator by the conjugate.

That is,

$$LHS = \frac{\cos\theta}{1 + \sin\theta} = \frac{\cos\theta\,(1 - \sin\theta)}{(1 + \sin\theta)(1 - \sin\theta)} = \frac{\cos\theta\,(1 - \sin\theta)}{1 - \sin^2\theta}$$

$$= \frac{\cos\theta\,(1 - \sin\theta)}{\cos^2\theta} = \frac{(1 - \sin\theta)}{\cos\theta} = RHS$$

The proof can also be executed by taking the RHS and converting it into the LHS.

$$RHS = \frac{1 - \sin\theta}{\cos\theta}$$

Multiplying by the conjugate of the numerator gives,

$$RHS = \frac{1 - \sin\theta}{\cos\theta} = \frac{(1 - \sin\theta)(1 + \sin\theta)}{\cos\theta\,(1 + \sin\theta)} = \frac{1 - \sin^2\theta}{\cos\theta\,(1 + \sin\theta)}$$

$$= \frac{\cos^2\theta}{\cos\theta\,(1 + \sin\theta)} = \frac{\cos\theta}{1 + \sin\theta} = LHS$$

The conjugate multiplication is also useful to prove some identities involving square roots.

<u>Example 8</u>: **Prove that**

$$\sqrt{\frac{1 + \sin\theta}{1 - \sin\theta}} = \sec\theta + \tan\theta$$

<u>Solution</u>: **Here,**

$$\text{LHS} = \sqrt{\frac{1 + \sin\theta}{1 - \sin\theta}}$$

Here the expression $\dfrac{1+\sin\theta}{1-\sin\theta}$ is inside the square root. So, let's multiply the expression within the square root by the conjugate of its denominator. That is,

$$\text{LHS} = \sqrt{\frac{1 + \sin\theta}{1 - \sin\theta}} = \sqrt{\frac{(1 + \sin\theta)(1 + \sin\theta)}{(1 - \sin\theta)(1 + \sin\theta)}}$$

$$= \sqrt{\frac{(1 + \sin\theta)^2}{1 - \sin^2\theta}} = \sqrt{\frac{(1 + \sin\theta)^2}{\cos^2\theta}} = \frac{\sqrt{(1 + \sin\theta)^2}}{\sqrt{\cos^2\theta}} = \frac{1 + \sin\theta}{\cos\theta}$$

$$= \frac{1}{\cos\theta} + \frac{\sin\theta}{\cos\theta} = \sec\theta + \tan\theta = \text{RHS}$$

The proof is also possible by multiplying with the conjugate of the numerator. That is,

$$\text{LHS} = \sqrt{\frac{(1 + \sin\theta)(1 - \sin\theta)}{(1 - \sin\theta)(1 - \sin\theta)}} = \sqrt{\frac{1 - \sin^2\theta}{(1 - \sin\theta)^2}}$$

$$= \sqrt{\frac{\cos^2\theta}{(1 - \sin\theta)^2}} = \frac{\cos\theta}{1 - \sin\theta}$$

Again, by conjugate multiplication,

$$\text{LHS} = \frac{\cos\theta}{1 - \sin\theta} = \frac{\cos\theta\,(1 + \sin\theta)}{(1 - \sin\theta)(1 + \sin\theta)}$$

$$= \frac{\cos\theta\,(1 + \sin\theta)}{1 - \sin^2\theta} = \frac{\cos\theta\,(1 + \sin\theta)}{\cos^2\theta}$$

$$= \frac{1 + \sin\theta}{\cos\theta} = \frac{1}{\cos\theta} + \frac{\sin\theta}{\cos\theta}$$

$$= \sec\theta + \tan\theta = \text{RHS}$$

<u>Problem 12</u>: **Prove the following identities.**

i. $\sqrt{\dfrac{1-\sin\theta}{1+\sin\theta}} = \sec\theta - \tan\theta$ ii. $\sqrt{\dfrac{\sec\theta-\tan\theta}{\sec\theta+\tan\theta}} = \dfrac{1}{\sec\theta+\tan\theta}$

iii. $\dfrac{1}{\sec\theta+\tan\theta} = \sec\theta - \tan\theta$ iv. $\dfrac{1-\cos\theta}{1+\cos\theta} = (\text{cosec}\,\theta - \cot\theta)^2$

v. $\dfrac{\sec\theta+\text{cosec}\,\theta}{\tan\theta+\cot\theta} = \sin\theta + \cos\theta$ vi. $\dfrac{1-2\cos^2\theta}{\sin\theta\cos\theta} = \tan\theta - \cot\theta$

Day Three

Session One
Unit circle and the
scope of trigonometry

Session Two
Quadrants and their sign convention

Session Three
Standard angles in the first quadrant

Meditation is like a gym in which you
develop the powerful mental muscle of
calm and insight.

- Ajahn Brahm

SESSION ONE

UNIT CIRCLE AND THE SCOPE OF TRIGONOMETRY

On day one, we saw that a rotating line segment with one end fixed will form a circle with the line segment as its radius, and the arc of rotation as its circumference.

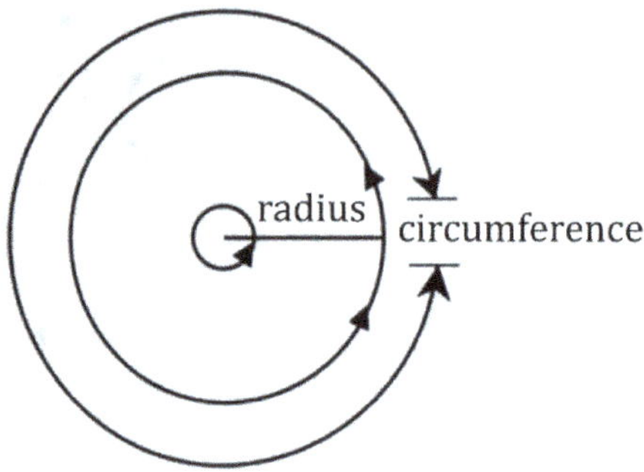

Today, we are going to draw this circle on a coordinate system. If you are not familiar with a coordinate system, see the box below. If you are, then skip it.

HOW TO WORK WITH COORDINATE SYSTEMS

A coordinate system is a device used to locate and name points in a plane. It consists of two infinitely long perpendicular lines called the 'X' and 'Y' axes, meeting at a single point called the 'origin'. Every point on the axes corresponds to a real number. The origin corresponds to the number zero.

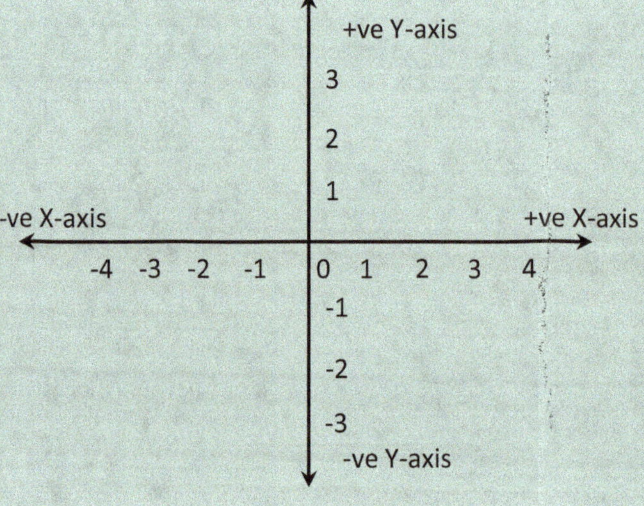

A point is named as a pair of numbers (x, y) called coordinates. The name (x, y) suggests that to reach the point, one has to move 'x' units along the direction of the X-axis, and then 'y' units along the direction of the Y-axis.

For instance, in the figure, the point P is named (3,2), since to reach P, we have to move 3 units along the direction of the positive X-axis and then 2 units along the direction of the positive Y-axis. The number 3 is called the 'x-coordinate' and 2 is called the 'y-coordinate'.

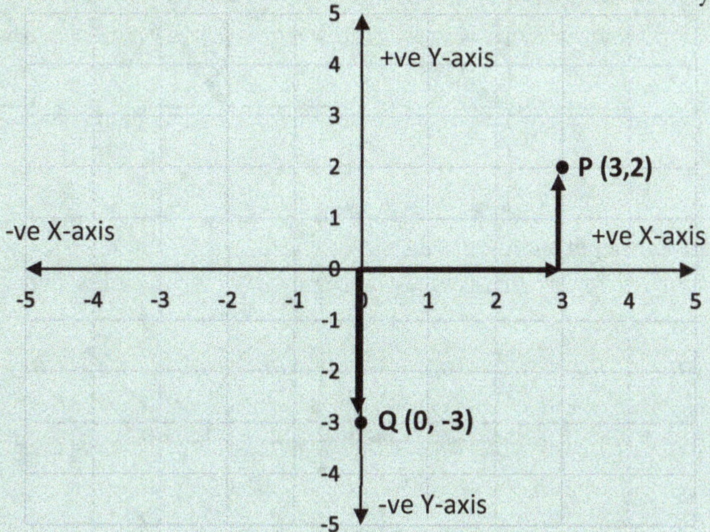

The point Q is named $(0, -3)$, since to reach Q we have to move 0 units along the X-axis and then 3 units along the negative Y-axis. Here 0 is the x-coordinate and -3 is the y-coordinate.

For all points along the X-axis, the y-coordinate is zero, and for all points along the Y-axis, the x-coordinate is zero.

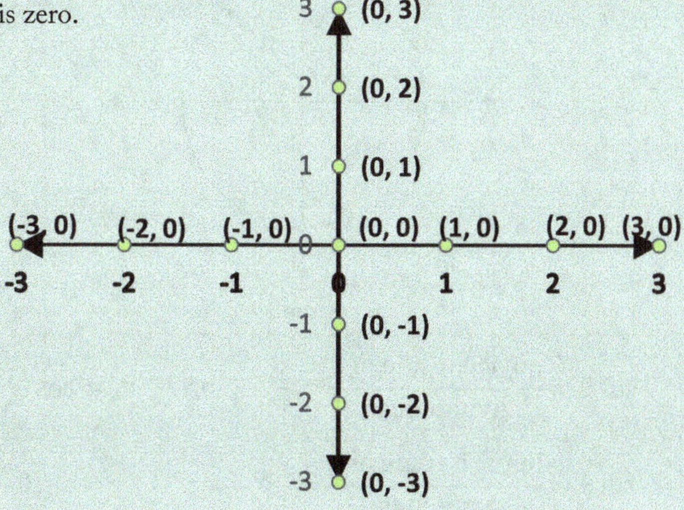

Problem: Draw a coordinate system and locate the following points:

i. $(0, -1.5)$ **ii.** $(-4, -3)$ **iii.** $(1, -2)$ **iv.** $(6, 4)$
v. $(-4, 3)$ **vi.** $(1, 1.5)$

Consider a line segment of unit length placed on the positive X-axis with one end fixed at the origin and the other end free to rotate counter clockwise (see figure 24).

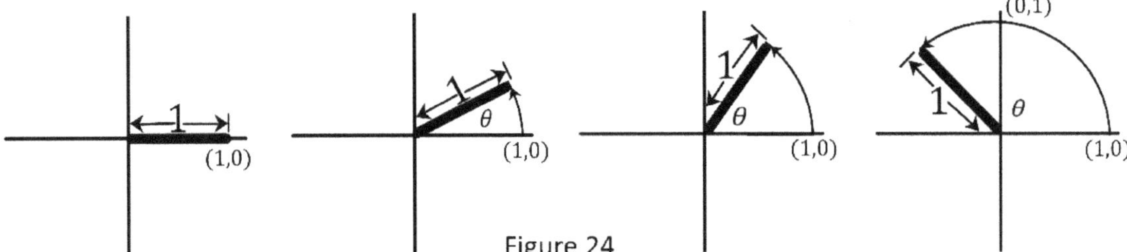

Figure 24

After a cycle of rotation, we get a circle,

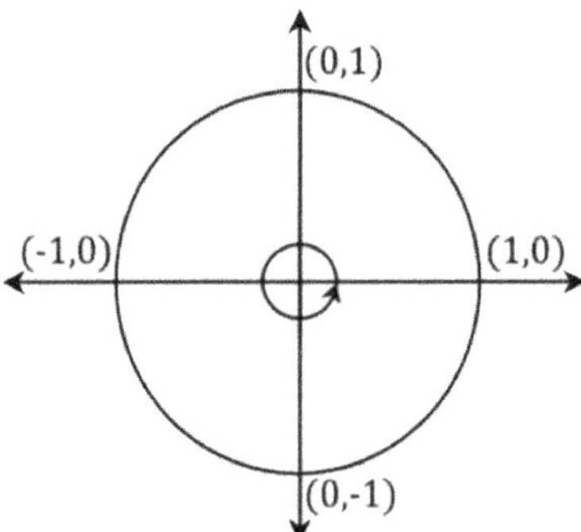

Let (x, y) be a point on the circle, and $'\theta'$ be the angle of rotation of the radius up to that point. (see figure 25).

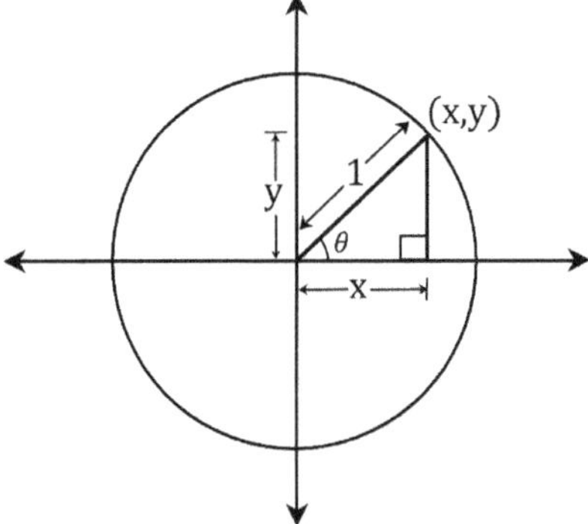

Figure 25

Such that,

$$\cos\theta = \frac{\text{adjacent side of } \theta}{\text{hypotenuse}} = \frac{x}{1} \qquad \text{or,} \qquad x = \cos\theta \qquad \text{--- (28)}$$

$$\sin\theta = \frac{\text{opposite side of } \theta}{\text{hypotenuse}} = \frac{y}{1} \qquad \text{or,} \qquad y = \sin\theta \qquad \text{--- (29)}$$

That is,

$$x = \cos\theta$$
$$y = \sin\theta$$

These equations are known as the **parametric equations** of a circle, since they express every point (x, y) on the circle in terms of a single parameter – 'The angle of rotation (θ)'.

$$(x, y) = (\cos\theta, \sin\theta)$$

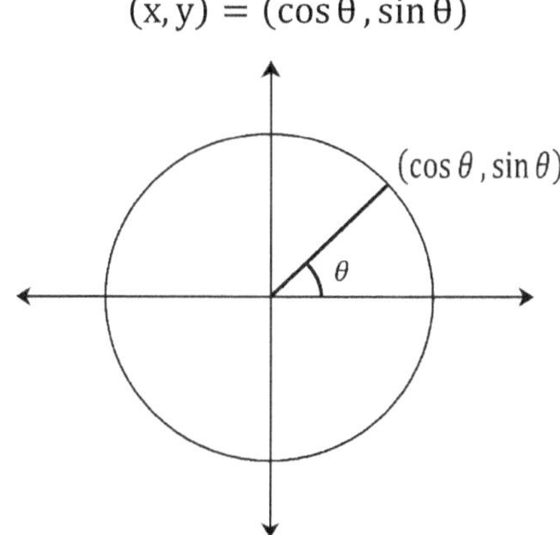

Starting from $0°$, the angle of rotation increases, and as the rotation progresses it becomes $30°, 45°, 60°, 90°, 180°, 270°, 360°$ and so on.

Corresponding to each value of θ, the coordinates become $(\cos\theta, \sin\theta)$.

$$\text{when, } \theta = 0° \ ; \ (x, y) = (\cos 0°, \sin 0°)$$

$$\text{when, } \theta = 30° \ ; \ (x, y) = (\cos 30°, \sin 30°)$$

$$\text{when, } \theta = 90° \ ; \ (x, y) = (\cos 90°, \sin 90°) \text{ and so on.}$$

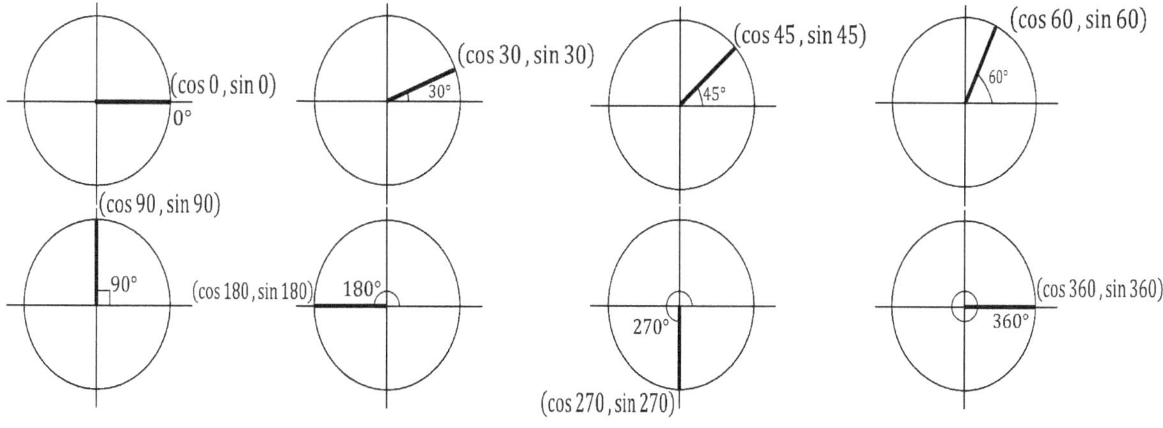

Notice that the trigonometric functions like $\cos 180°$, $\sin 270°$, etc are not possible on a right-angled triangle since those angles are impossible inside a triangle.

But, using the unit circle and its parametric equations, trigonometric functions can be defined for angles greater than 90°, thereby expanding the scope and applications of trigonometry into much larger domains.

An immediate application is to find the values of the trigonometric functions for the quadrant angles 0°, 90°, 180°, 270° and 360°.

The method is surprisingly simple

See the figure 26,

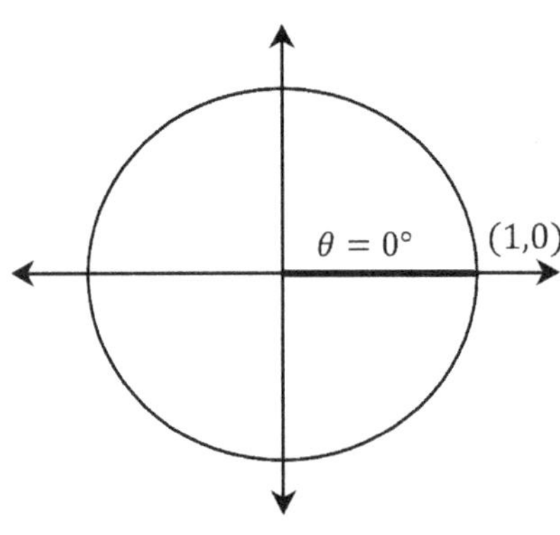

Figure 26

$$\text{when, } \theta = 0° \; ; \; (x, y) = (\cos 0°, \sin 0°) = (1,0)$$

$$\text{which gives, } \sin 0° = 0 \text{ and } \cos 0° = 1$$

$$\text{thus, } \tan 0° = \frac{\sin 0°}{\cos 0°} = \frac{0}{1} = 0$$

$$\operatorname{cosec} 0° = \frac{1}{\sin 0°} = \frac{1}{0}$$

Since division by zero is not defined in mathematics, cosec 0° is not defined (see next page).

$$\sec 0° = \frac{1}{\cos 0°} = \frac{1}{1} = 1$$

$$\cot 0° = \frac{\cos 0°}{\sin 0°} = \frac{1}{0}, \text{which is also not defined.}$$

To summarise,

sin 0° = 0	**cosec 0° = Not defined (N. D)**
cos 0° = 1	**sec 0° = 1**
tan 0° = 0	**cot 0° = Not defined (N. D)**

DIVISION BY ZERO

Division by zero is not defined in standard mathematics. To demonstrate this, let's divide the number 2 by zero.

$$\frac{2}{0}$$

Assume that this division is possible and an answer exists.

Let it be equal to 'x'. i.e.

$$\frac{2}{0} = x$$

Which gives,

$$2 = x(0)$$

Meaning, x times 0 equals 2.

No x can satisfy this equation since multiplication by zero always yields zero.

This shows that the number x cannot exist and the division $\frac{2}{0} = x$ is meaningless and therefore cannot be defined.

when, $\theta = 90°$; $(x, y) = (\cos 90°, \sin 90°) = (0,1)$ (see figure 27)

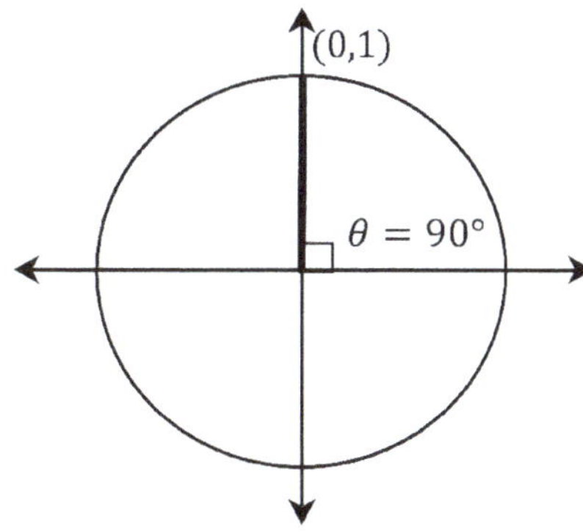

Figure 27

which gives, $\sin 90° = 1$ and $\cos 90° = 0$

thus, $\tan 90° = \dfrac{\sin 90°}{\cos 90°} = \dfrac{1}{0} = $ Not Defined (N. D)

$$\operatorname{cosec} 90° = \frac{1}{\sin 90°} = \frac{1}{1} = 1$$

$$\sec 90° = \frac{1}{\cos 90°} = \frac{1}{0} = \text{N. D}$$

$$\cot 90° = \frac{\cos 90°}{\sin 90°} = \frac{0}{1} = 0$$

To summarise,

$\sin 90° = 1$	$\operatorname{cosec} 90° = 1$
$\cos 90° = 0$	$\sec 90° = \text{N. D}$
$\tan 90° = \text{N. D}$	$\cot 90° = 0$

Problem 13: **Complete the empty columns in the table showing trigonometric functions of the quadrant angles 0°, 90°, 180°, 270° and 360°.**

	$\theta = 0°$	$\theta = 90°$	$\theta = 180°$	$\theta = 270°$	$\theta = 360°$
$\sin \theta$	0	1			
$\cos \theta$	1	0			
$\tan \theta$	0	N. D			
$\operatorname{cosec} \theta$	N. D	1			
$\sec \theta$	1	N. D			
$\cot \theta$	N. D	0			

Memorising the entire table is a real mind job and a waste of time. If you have sine and cosine in your pocket, the rest can be easily deduced. The sine and cosine of an angle is given by the parametric equations,

$$x = \cos \theta$$
$$y = \sin \theta$$

applied to the coordinates corresponding to that angle.

So, you must remember these equations.

Try the code name: **X-BOX**, and connect it with the term **X-COS**, which corresponds to the equation

$$x = \cos \theta.$$

If you remember $x = \cos \theta$, it's obvious that $y = \sin \theta$.

SESSION TWO

QUADRANTS AND THEIR SIGN CONVENTION

The X and Y axes divide the plain into four sectors called **quadrants**. They are named as I, II, III and IV in the order of increasing angle of rotation (see figure 28).

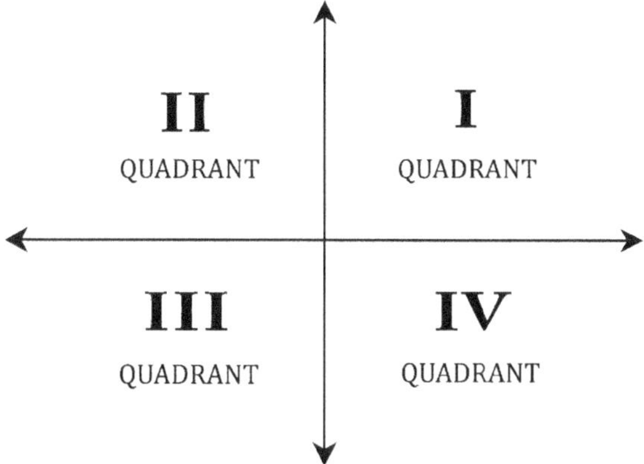

Figure 28

Depending upon the signs of the coordinate axes, different trigonometric functions have different signs in each quadrant. Since much of the calculations in the upcoming sessions depend on this, we dedicate this session to discuss the quadrants and their sign convention. There are no problems in this session, nor are there any examples. For some, it might feel boring, but bear with me, it's important. If you are in a hurry, at least see the summary boxes and memorise the shortcut phrase at the end.

I- QUADRANT

In the first quadrant both 'x' and 'y' coordinates are positive (see figure 29).

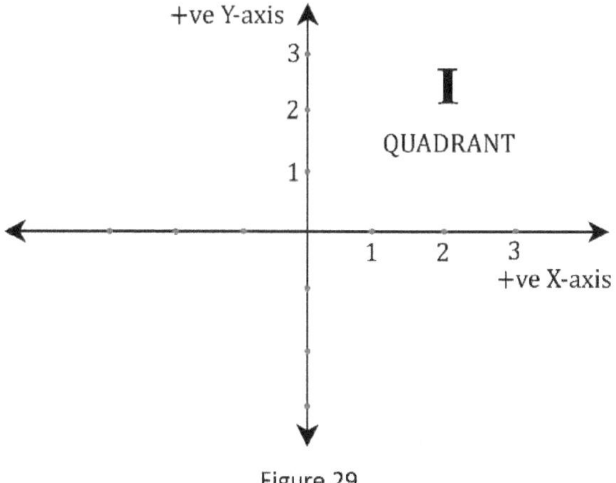

Figure 29

Since the values of the coordinates 'x' and 'y' gives the functions cos θ and sin θ,

$$x = \cos \theta$$

$$y = \sin \theta$$

the functions sin θ and cos θ also become positive (see figure 30).

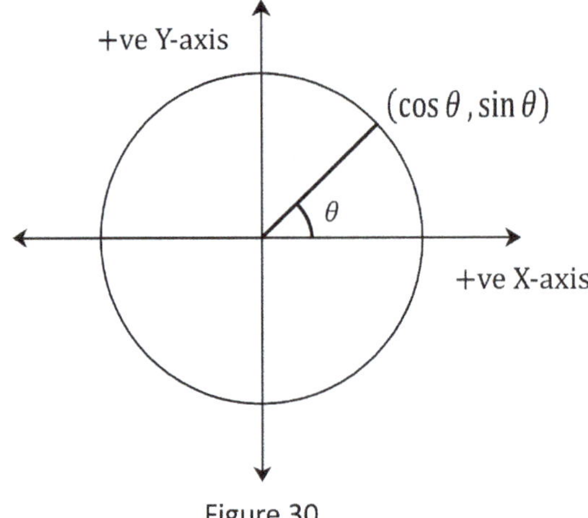

Figure 30

And since all the other functions are the ratios and reciprocals of sin θ and cos θ, they are also positive. To summarise,

> **In the I- quadrant, ALL trigonometric functions are positive**

II- QUADRANT

In the second quadrant, the Y-axis is positive, but the X-axis is negative (see figure 31).

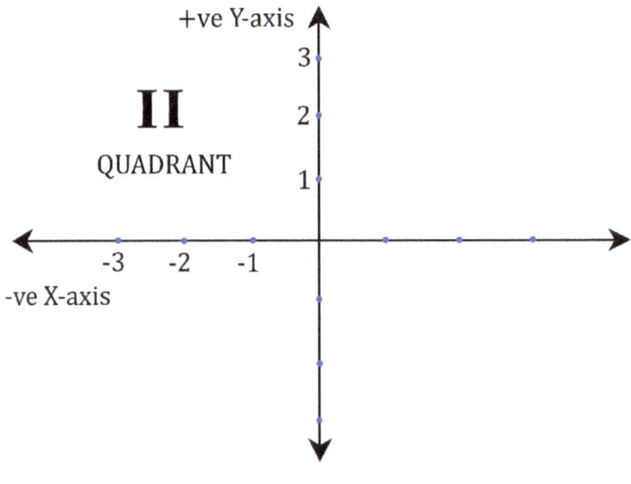

Figure 31

Hence, sin θ is positive, and cos θ is negative (see figure 32).

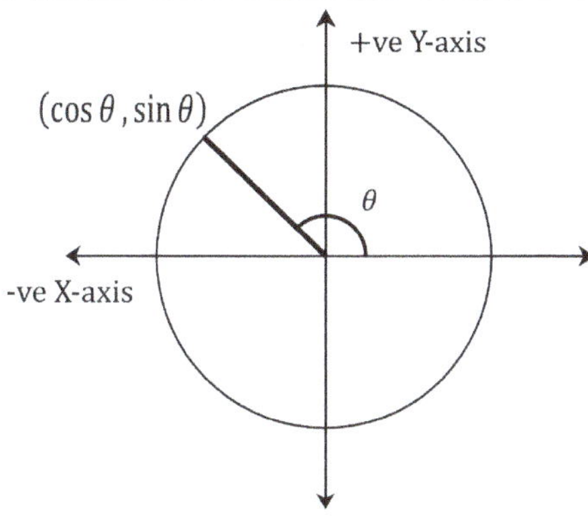

Figure 32

Since $\sin\theta$ is positive, its reciprocal $\csc\theta$ is also positive, and since $\cos\theta$ is negative its reciprocal $\sec\theta$ is negative.

$$\csc\theta = \frac{1}{\sin\theta} = \frac{1}{+ve} = +ve$$

$$\sec\theta = \frac{1}{\cos\theta} = \frac{1}{-ve} = -ve$$

And since $\cos\theta$ is negative, both $\tan\theta$ and $\cot\theta$ are also negative.

$$\tan\theta = \frac{\sin\theta}{\cos\theta} = \frac{+ve}{-ve} = -ve$$

$$\cot\theta = \frac{\cos\theta}{\sin\theta} = \frac{-ve}{+ve} = -ve$$

To summarise,

In the II - quadrant, only SINE and its RECIPROCAL are positive

III- QUADRANT

In the third quadrant, both 'x' and 'y' are negative (see figure 33).

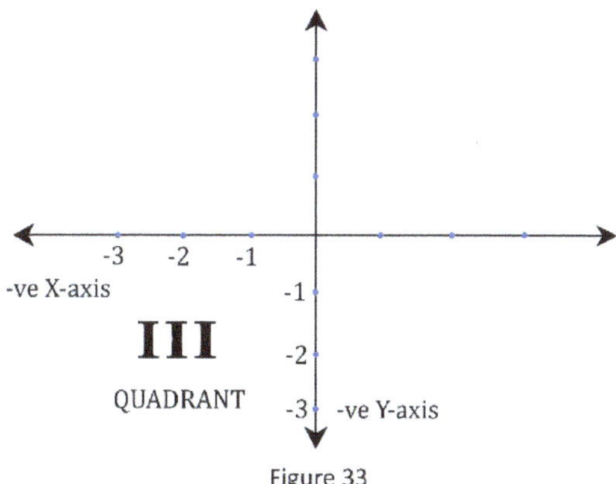

Figure 33

Here both $\sin\theta$ and $\cos\theta$ and are negative (see figure 34).

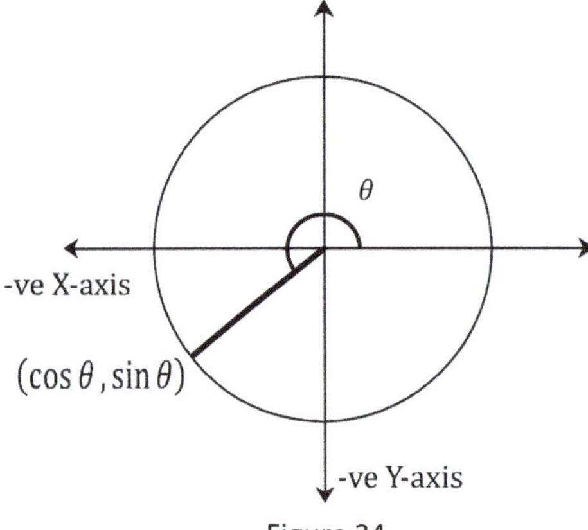

Figure 34

Since $\sin\theta$ and $\cos\theta$ are negative, their reciprocals $\operatorname{cosec}\theta$ and $\sec\theta$ are also negative.

$$\operatorname{cosec}\theta = \frac{1}{\sin\theta} = \frac{1}{-ve} = -ve$$

$$\sec\theta = \frac{1}{\cos\theta} = \frac{1}{-ve} = -ve$$

And since $\sin\theta$ and $\cos\theta$ are negative, $\tan\theta$ and its reciprocal $\cot\theta$ are positive.

$$\tan\theta = \frac{\sin\theta}{\cos\theta} = \frac{-ve}{-ve} = +ve$$

$$\cot\theta = \frac{\cos\theta}{\sin\theta} = \frac{-ve}{-ve} = +ve$$

To summarise,

In the **III**-quadrant, only **TAN** and its **RECIPROCAL** are positive

IV- QUADRANT

In the fourth quadrant, the X-axis is positive, but the Y-axis is negative (see figure 35).

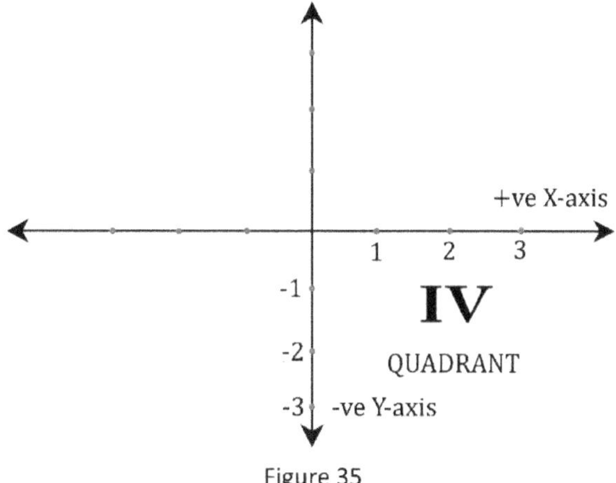

Figure 35

Hence, $\cos\theta$ is positive, and $\sin\theta$ is negative (see figure 36).

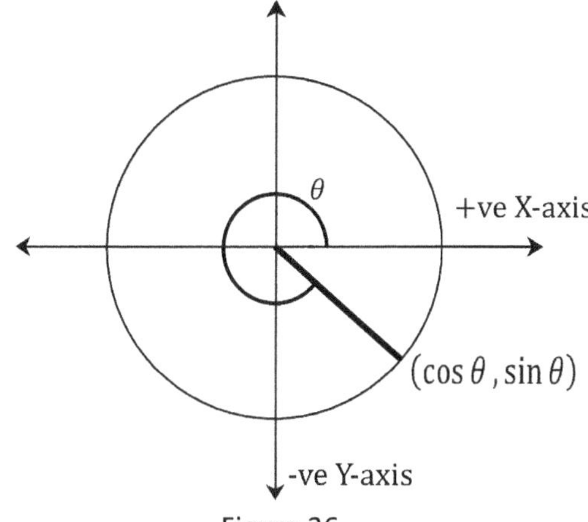

Figure 36

Since $\cos\theta$ is positive, its reciprocal $\sec\theta$ is also positive, and since $\sin\theta$ is negative, its reciprocal $\csc\theta$ is negative.

$$\sec\theta = \frac{1}{\cos\theta} = \frac{1}{+ve} = +ve$$

$$\csc\theta = \frac{1}{\sin\theta} = \frac{1}{-ve} = -ve$$

And since $\sin\theta$ is negative, both $\tan\theta$ and $\cot\theta$ are also negative.

$$\tan\theta = \frac{\sin\theta}{\cos\theta} = \frac{-ve}{+ve} = -ve$$

$$\cot\theta = \frac{\cos\theta}{\sin\theta} = \frac{+ve}{-ve} = -ve$$

To summarise,

> ## In the IV -quadrant, only COS and its RECIPROCAL are positive

It is important to know which functions are positive in each quadrant. Many of the calculations in the upcoming sessions crucially depends on this knowledge.

Memorise the phrase,

All Silver Tea Cup

All → **A**ll functions are positive in the I-Quadrant

Silver → **S**ine and its reciprocal are positive in the II-Quadrant

Tea → **T**an and its reciprocal are positive in the III-Quadrant

Cup → **C**os and its reciprocal are positive in the IV-Quadrant

SESSION THREE

STANDARD ANGLES IN THE FIRST QUADRANT

In this session, we shall look at some angles which occur frequently in trigonometry. They are 30°, 45° and 60°. All of them lie in the first quadrant.

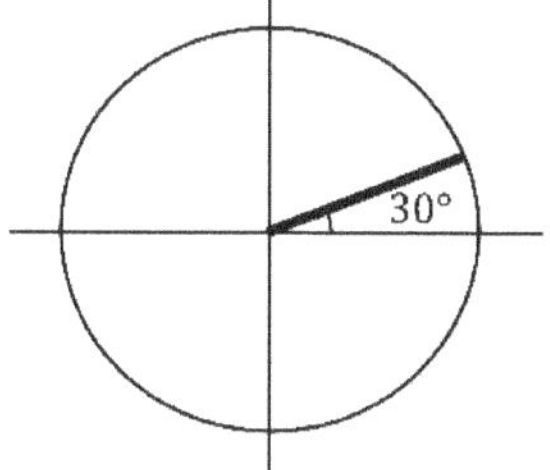
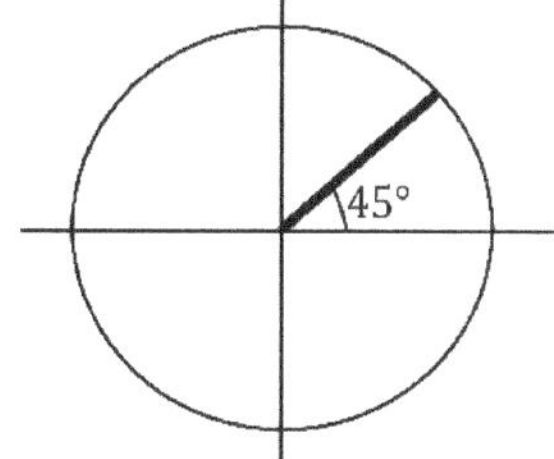
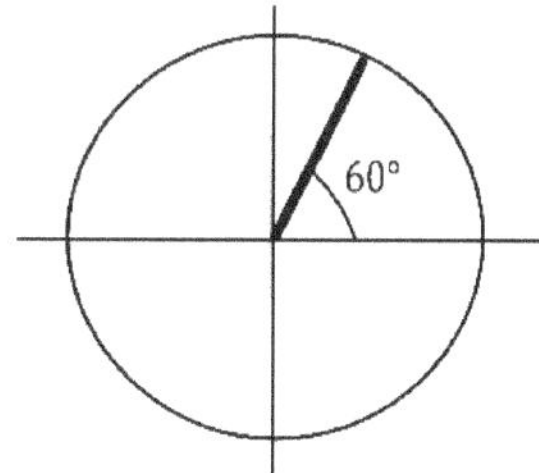

First, let's consider 45°.

45°

(If you are not interested in the proof, skip it and learn the equation (30) and problem 14.)

At 45°, the radius cuts the first quadrant of the unit circle into two exact halves, such that the 'x' coordinate and 'y' coordinate become equal (see figure 37).

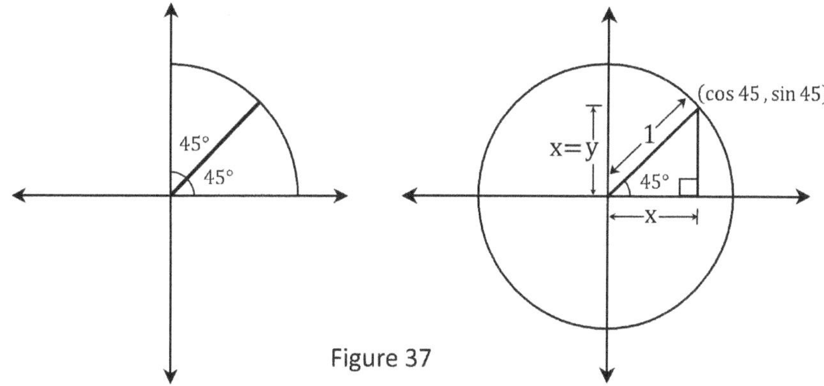

Figure 37

The equality of 'x' and 'y' coordinates implies the equality of $\cos\theta$ and $\sin\theta$. That is, at 45°

$$x = y$$

$$\text{or,} \quad \cos\theta = \sin\theta$$

Therefore, the identity $\sin^2\theta + \cos^2\theta = 1$ becomes,

$$\sin^2 45° + \cos^2 45° = \sin^2 45° + \sin^2 45° = 2\sin^2 45° = 1$$

$$\text{or,} \quad \sin 45° = \pm\frac{1}{\sqrt{2}}$$

Since all trigonometric functions are positive in the first quadrant (**A**ll **S**ilver **T**ea **C**up),

$$\sin 45° = \frac{1}{\sqrt{2}}$$

And since $\sin 45° = \cos 45°$, we get,

$$\cos 45° = \frac{1}{\sqrt{2}}$$

To summarise,

$$\boxed{\sin 45° = \cos 45° = \frac{1}{\sqrt{2}}}$$

--- (30)

From $\sin\theta$ and $\cos\theta$ all the other functions can be calculated. See problem 14.

<u>**Problem 14**</u>: **Complete the table containing the trigonometric functions of 45°.**

	$\theta = 45°$
$\sin\theta$	$\dfrac{1}{\sqrt{2}}$
$\cos\theta$	$\dfrac{1}{\sqrt{2}}$
$\tan\theta$	
$\operatorname{cosec}\theta$	
$\sec\theta$	
$\cot\theta$	

<u>60° and 30°</u>

(If you are not interested in the proof, skip it and learn the equation, (31), (32) and problem 15.)

Let ABC be an equilateral triangle of unit side length (See figure 38).

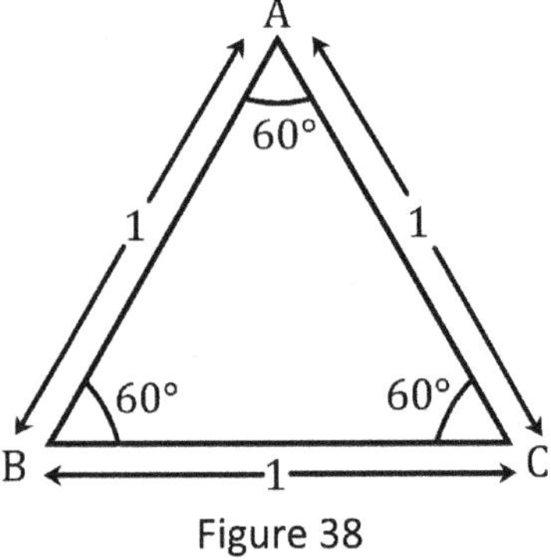

Figure 38

Draw an angle bisector at A such that it bisects both angle A and the side BC into two exact halves (see figure 39).

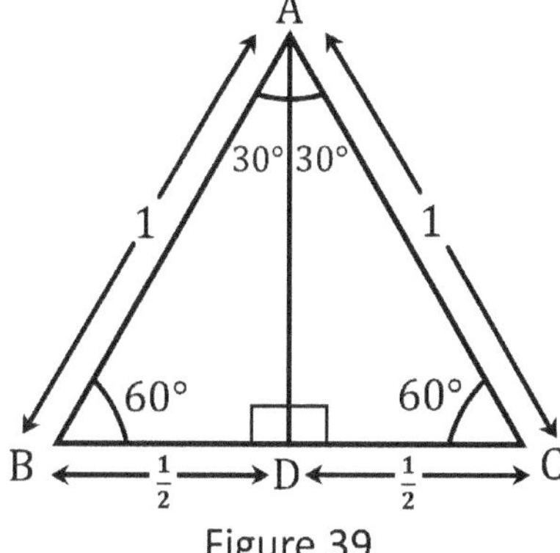

Figure 39

The bisector AD divides ABC into two right-angled triangles ABD and ACD with angles 30° and 60° each. Place the triangle on the unit circle such that the side BC coincides with the X-axis (see figure 40).

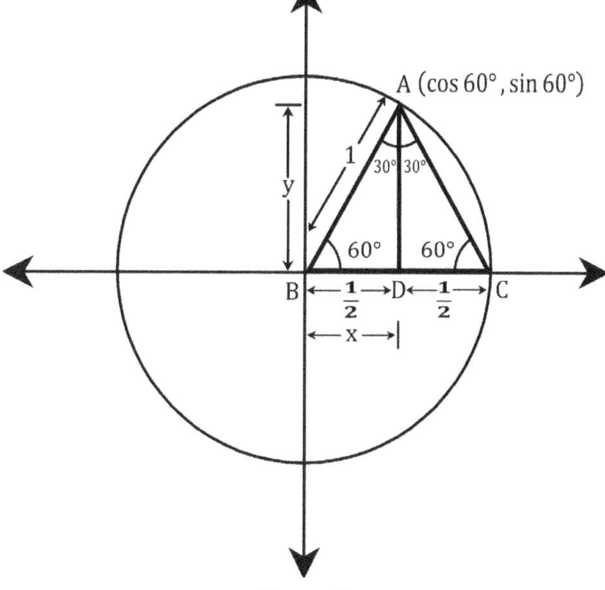

Figure 40

Here the coordinates (x, y) of the point A is $(\cos 60°, \sin 60°)$, therefore,

$$\cos 60° = x = \frac{1}{2}$$

And using the identity $\sin^2 \theta + \cos^2 \theta = 1$ we get,

$$\sin^2 60° + \cos^2 60° = \sin^2 60° + \frac{1}{4} = 1$$

$$\text{i.e.} \quad \sin^2 60° = \frac{3}{4}$$

$$\text{or,} \quad \sin 60° = \pm \frac{\sqrt{3}}{2}$$

Since 60° is in the first quadrant,

$$\sin 60° = \frac{\sqrt{3}}{2}$$

To summarise, we have,

$$\boxed{\sin 60° = \frac{\sqrt{3}}{2} \ , \ \cos 60° = \frac{1}{2}} \qquad \text{--- (31)}$$

From triangle ABD, we get

$$\sin 30° = \frac{\text{opposite side}}{\text{hypotenuse}} = \frac{\frac{1}{2}}{1} = \frac{1}{2}$$

$$\text{Therefore,} \quad \sin^2 30° + \cos^2 30° = \frac{1}{4} + \cos^2 30° = 1$$

$$\text{i.e.} \quad \cos 30° = \pm \frac{\sqrt{3}}{2}$$

Since 30° is in the first quadrant, we get,

$$\cos 30° = \frac{\sqrt{3}}{2}$$

To summarise, we have,

$$\boxed{\sin 30° = \frac{1}{2} \ , \ \cos 30° = \frac{\sqrt{3}}{2}} \qquad \text{--- (32)}$$

Problem 15: **Complete the table containing the trigonometric functions of 30° and 60°**

	$\theta = 30°$	$\theta = 60°$
$\sin \theta$	$\dfrac{1}{2}$	$\dfrac{\sqrt{3}}{2}$
$\cos \theta$	$\dfrac{\sqrt{3}}{2}$	$\dfrac{1}{2}$
$\tan \theta$		
$\operatorname{cosec} \theta$		
$\sec \theta$		
$\cot \theta$		

If it seems a lot, at least try to remember sine and cosine. From those two, others can be easily deduced.

Memorize the numbers,

$$\frac{1}{2}, \frac{1}{\sqrt{2}} \text{ and } \frac{\sqrt{3}}{2} \text{ in order.}$$

They are the sine values of 30°, 45°, and 60° respectively. If you have those, finding cosine is easy. See the table below

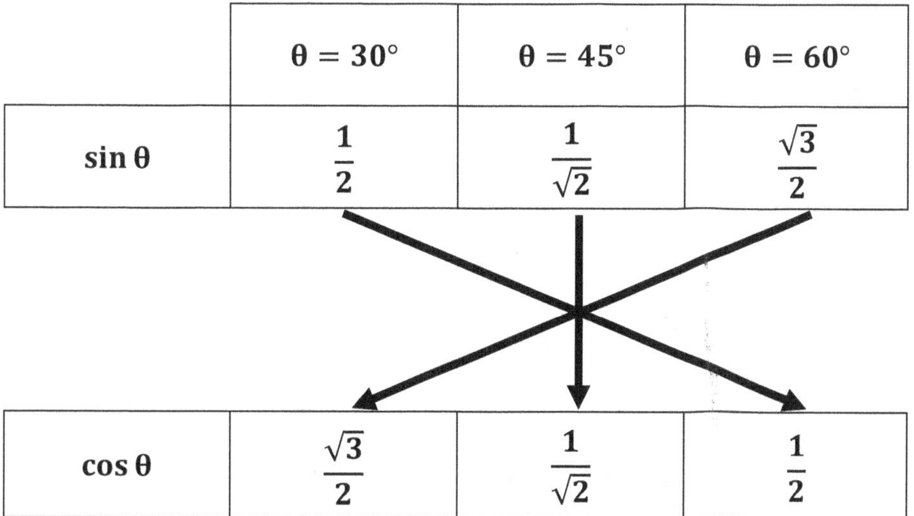

	$\theta = 30°$	$\theta = 45°$	$\theta = 60°$
$\sin \theta$	$\dfrac{1}{2}$	$\dfrac{1}{\sqrt{2}}$	$\dfrac{\sqrt{3}}{2}$
$\cos \theta$	$\dfrac{\sqrt{3}}{2}$	$\dfrac{1}{\sqrt{2}}$	$\dfrac{1}{2}$

Before calling it a day, meditate, relax and recall what we have studied. These concepts are central and crucial to trigonometry. It's essential that you have a clear and crisp understanding of them. If you could not complete the sessions in one day, don't worry, it's ok. Tomorrow, return to the same sessions. Try to create a good mental picture and make flash cards for future revisions.

Day Four

Session One
Problem session

Session Two
Negative angles
Distance formula

Session Three
Trigonometric ratios of sum and difference

A man who dares to waste one hour of time has not discovered the value of life.

- Charles Darwin

SESSION ONE

PROBLEM SESSION

Solving problems are a crucial part of mathematical training. The more you work on problems, the more you become an expert. So, let's dedicate this session to solve some problems. But first, take a moment to meditate and recollect what we have learned yesterday. Recall the sign convention of the quadrants, the trigonometric functions of the quadrant angles and the standard angles in the first quadrant.

Example 9: Find the value of

$$\sin\frac{\pi}{3}\cos\frac{\pi}{6} + \cos\frac{\pi}{3}\sin\frac{\pi}{6}$$

Solution: We know that,

$$\frac{\pi}{3}\,\text{rad} = 60° \text{ and } \frac{\pi}{6}\,\text{rad} = 30°$$

Therefore,

$$\sin\frac{\pi}{3}\cos\frac{\pi}{6} + \cos\frac{\pi}{3}\sin\frac{\pi}{6} = \sin 60° \cos 30° + \cos 60° \sin 30°$$

$$\left(\frac{\sqrt{3}}{2}\right)\left(\frac{\sqrt{3}}{2}\right) + \left(\frac{1}{2}\right)\left(\frac{1}{2}\right) = \frac{3}{4} + \frac{1}{4} = 1$$

Problem 16: Find the values of the following.

i. $\quad 4\sin^3\frac{\pi}{3} - 3\cos\frac{\pi}{6}$

ii. $\quad \tan^2 60° + \tan^2 30°$

iii. $\quad \sin^3\frac{\pi}{3}\cos^2\frac{\pi}{4}\tan\frac{\pi}{6}$

iv. $\quad \sin\frac{\pi}{2}\cos\frac{\pi}{4}\tan\frac{\pi}{6}$

Ans:

i. $\quad 0 \qquad$ ii. $\quad \dfrac{10}{3} \qquad$ iii. $\quad \dfrac{3}{16} \qquad$ iv. $\quad \dfrac{1}{\sqrt{6}}$

Example 10: Prove that

$$\frac{\tan 45^\circ - \tan 30^\circ}{1 + \tan 45^\circ \tan 30^\circ} = 2 - \sqrt{3}$$

Solution: Here,

$$\text{LHS} = \frac{\tan 45^\circ - \tan 30^\circ}{1 + \tan 45^\circ \tan 30^\circ} = \frac{1 - \dfrac{1}{\sqrt{3}}}{1 + 1\left(\dfrac{1}{\sqrt{3}}\right)} = \frac{\dfrac{\sqrt{3}-1}{\sqrt{3}}}{\dfrac{\sqrt{3}+1}{\sqrt{3}}} = \frac{\sqrt{3}-1}{\sqrt{3}+1}$$

By conjugate multiplication,

$$\text{LHS} = \frac{(\sqrt{3}-1)(\sqrt{3}-1)}{(\sqrt{3}+1)(\sqrt{3}-1)} = \frac{(\sqrt{3}-1)^2}{(3-1)} = \frac{3 - 2\sqrt{3} + 1}{2} = \frac{4 - 2\sqrt{3}}{2}$$

$$\frac{2(2-\sqrt{3})}{2} = (2 - \sqrt{3}) = \text{RHS}$$

Problem 17: Prove the following.

i. $\dfrac{\tan 60^\circ - \tan 45^\circ}{1 + \tan 60^\circ \tan 45^\circ} = 2 - \sqrt{3}$

ii. $\cos\dfrac{\pi}{4}\cos\dfrac{\pi}{3} - \sin\dfrac{\pi}{4}\sin\dfrac{\pi}{3} = \dfrac{1-\sqrt{3}}{2\sqrt{2}}$

iii. $\operatorname{cosec}^2 45^\circ \cos^2 60^\circ \sec^2 30 = \dfrac{2}{3}$

iv. $4\tan^2 60^\circ - 3\tan^2 30^\circ + \tan^2 45^\circ = 12$

Example 11: If $\tan\theta = \dfrac{5}{12}$ and θ lies in the third quadrant, find all the other trigonometric functions.

Solution: Here, $\tan\theta = \dfrac{5}{12}$

By using the identity $1 + \tan^2\theta = \sec^2\theta$. We get,

$$\sec^2\theta = 1 + \frac{25}{144} = \frac{169}{144}$$

$$\text{i.e.} \quad \sec\theta = \pm\sqrt{\frac{169}{144}} = \pm\frac{13}{12}$$

Since the angle θ is in the third quadrant, $\sec\theta$ is negative. Therefore,

$$\sec\theta = -\frac{13}{12}$$

$$\text{and,} \quad \cos\theta = \frac{1}{\sec\theta} = \frac{1}{-\dfrac{13}{12}} = -\frac{12}{13}$$

By using the identity $\sin^2\theta + \cos^2\theta = 1$. We get,

$$\sin^2\theta + \left(-\frac{12}{13}\right)^2 = \sin^2\theta + \frac{144}{169} = 1$$

$$\text{i.e.} \quad \sin^2\theta = \frac{25}{169}$$

$$\text{or,} \quad \sin\theta = \pm\sqrt{\frac{25}{169}} = \pm\frac{5}{13}$$

Since θ is in the third quadrant, $\sin\theta$ is negative. That is,

$$\sin\theta = -\frac{5}{13}$$

$$\text{and,} \quad \text{cosec}\,\theta = \frac{1}{\sin\theta} = \frac{1}{-\dfrac{5}{13}} = -\frac{13}{5}$$

Since $\tan\theta = \dfrac{5}{12}$, we get,

$$\cot\theta = \frac{1}{\tan\theta} = \frac{1}{\dfrac{5}{12}} = \frac{12}{5}$$

To summarize, we have the trigonometric functions,

$$\sin\theta = -\frac{5}{13} \qquad \text{cosec}\,\theta = -\frac{13}{5}$$

$$\cos\theta = -\frac{12}{13} \qquad \sec\theta = -\frac{13}{12}$$

$$\tan\theta = \frac{5}{12} \qquad \cot\theta = \frac{12}{5}$$

Problem 18: Find all the trigonometric functions if:

i. $\tan\theta = \dfrac{24}{7}$ and θ lies in the third quadrant.

ii. $\sin\theta = \dfrac{1}{2}$ and θ lies in the first quadrant (θ is acute).

iii. $\cot\theta = \dfrac{24}{7}$ and θ is acute.

iv. $\sin\theta = -\dfrac{4}{5}$ and θ lies in the fourth quadrant.

v. $\sec\theta = -\dfrac{13}{12}$ and θ lies in the third quadrant.

vi.   If $\cos\theta = \dfrac{1}{2}$ and θ lies in the first quadrant, find $5\sin\theta + 3\tan\theta$.

SESSION TWO

NEGATIVE ANGLES

It is a convention in trigonometry to assign positive values to rotations in the counter-clockwise direction and negative values to rotations in the clockwise direction (see figure 41).

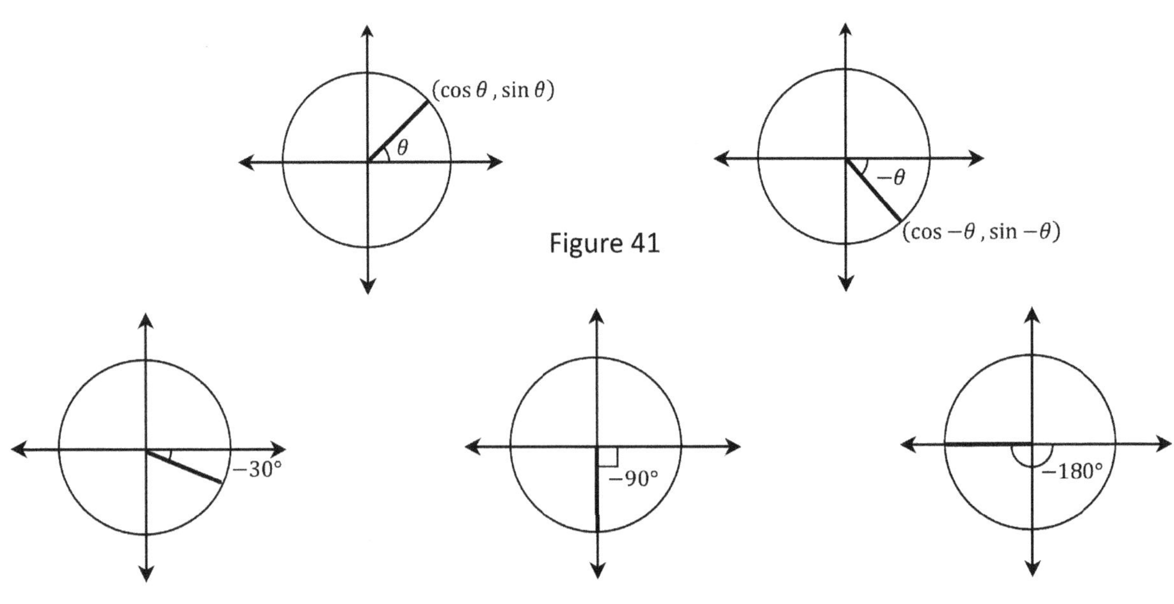

Figure 41

From the figure 42.

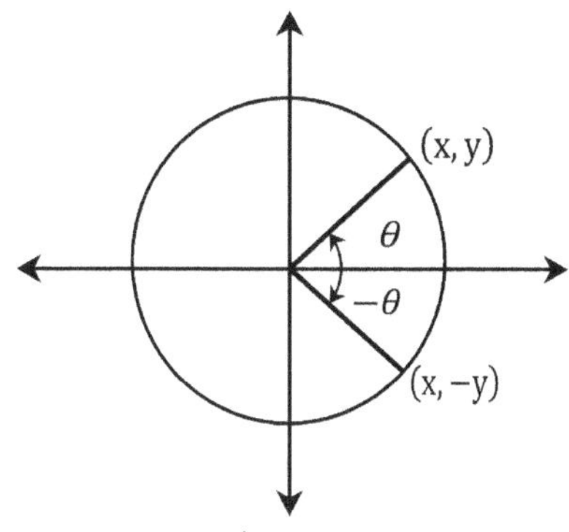

Figure 42

We can write,

$$\cos\theta = x \ \text{ and } \ \sin\theta = y \qquad\qquad \text{--- (33)}$$

$$\text{and } \ \cos(-\theta) = x \ \text{ and } \ \sin(-\theta) = -y \qquad\qquad \text{--- (34)}$$

From (33) and (34), we get,

$$\boxed{\begin{aligned}\cos(-\theta) &= \cos\theta \\ \sin(-\theta) &= -\sin\theta\end{aligned}}$$

From $\sin(-\theta)$ and $\cos(-\theta)$, we get,

$$\tan(-\theta) = \frac{\sin(-\theta)}{\cos(-\theta)} = -\frac{\sin\theta}{\cos\theta}$$

i.e.

$$\boxed{\tan(-\theta) = -\tan\theta}$$

Problem 19: **Find the values of the trigonometric functions of the following angles.**

    i. $\cos(-45°)$ ii. $\cos(-30°)$ iii. $\sin(-30°)$

    iv. $\cot(-60°)$ v. $\sin(-60°)$ vi. $\sin(-90°)$

Ans:

  i. $\dfrac{1}{\sqrt{2}}$ ii. $\dfrac{\sqrt{3}}{2}$ iii. $-\dfrac{1}{2}$ iv. $-\dfrac{1}{\sqrt{3}}$ v. $-\dfrac{\sqrt{3}}{2}$ vi. -1

Now, consider the term,

$$180° - 60°$$

Here $180°$ is positive and $60°$ is negative. So first, rotate the radius up to $180°$ in the counter-clockwise (positive) direction (see figure 43).

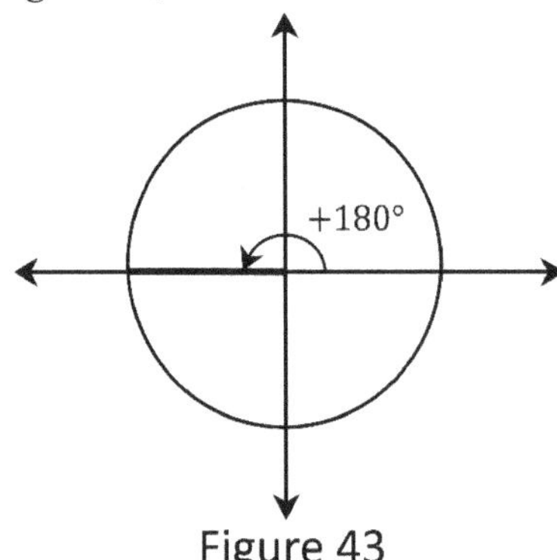

Figure 43

Then rotate $60°$ backwards in the clockwise (negative) direction to obtain,

$$180° - 60° = +120°(\text{see figure 44})$$

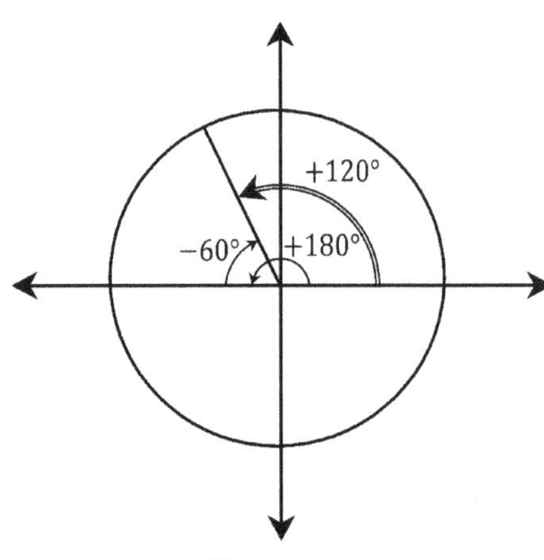

Figure 44

> ## <u>Problem 20</u>: Draw the rotations corresponding to the following angles.
>
> i. $30° - 60°$ ii. $60° + 30°$ iii. $180° - 20°$
>
> iv. $-270° - 30°$ v. $270° - 30°$

DISTANCE FORMULA

The application of Pythagoras theorem to coordinate geometry yields a powerful result called 'The distance formula'. It enables us to calculate the distance between two points in an x-y coordinate system.

(We present the distance formula to prove the addition and subtraction formulae in the next session. If you are not interested in the proof, skip it and learn the equations (38), (39), (44) and (45) in the next session)

Consider the points A and B with coordinates (x_1, y_1) and (x_2, y_2) (see figure 45).

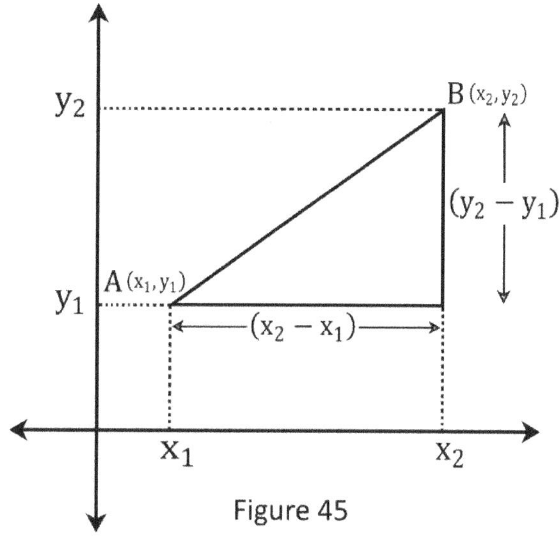

Figure 45

By Pythagoras theorem,

$$(AB)^2 = (x_2 - x_1)^2 + (y_2 - y_1)^2$$

$$\text{or, } \quad AB = \sqrt{(x_2 - x_1)^2 + (y_2 - y_1)^2} \qquad \text{--- (35)}$$

Problem 21: **Find the distance between the points A and B if:**

    i. $A = (0, 0)\,; B = (6, 8)$

    ii. $A = (4, 0)\,; B = (1, 4)$

    iii. $A = (-3, 2)\,; B = (3, 5)$

Ans:

    i. **10**   ii. **5**   iii. **$3\sqrt{5}$**

SESSION THREE

TRIGONOMETRIC FUNCTIONS OF SUM AND DIFFERENCE

(If you are not interested in the proof, skip it and learn the equations (38), (39), (44) and (45))

Let A and B be two angles on the unit circle with a difference A-B (see figure 46).

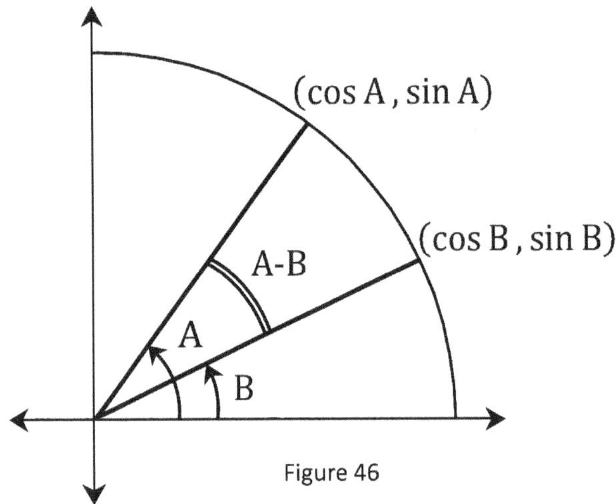

Figure 46

Let D be the distance between the points (cos A , sin A) and (cos B , sin B) corresponding to the angle $(A - B)$ (see figure 47).

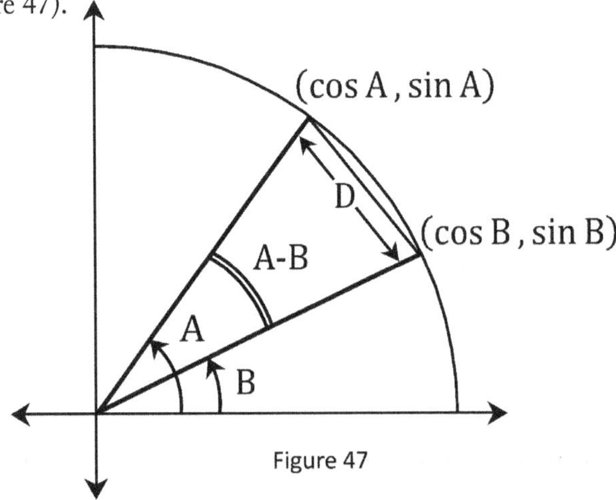

Figure 47

Using distance formula, we get,

$$D = \sqrt{(\cos A - \cos B)^2 + (\sin A - \sin B)^2}$$

or, $$D = \sqrt{\cos^2 A + \cos^2 B - 2\cos A \cos B + \sin^2 A + \sin^2 B - 2\sin A \sin B}$$

Since $\cos^2 A + \sin^2 A = \cos^2 B + \sin^2 B = 1$, we get,

$$D = \sqrt{2 - 2(\cos A \cos B + \sin A \sin B)}$$

or, $\quad D^2 = 2 - 2(\cos A \cos B + \sin A \sin B)$ $\qquad$ --- (36)

Now consider the angle of rotation A-B, starting from the positive X-axis (see figure 48).

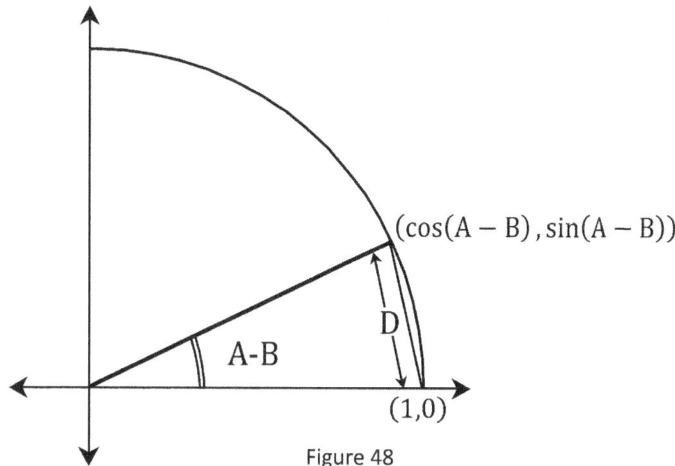

Figure 48

Since the angle A-B is same as the angle (A-B) from the previous figure (figure 47), the distance D remains constant. i.e.

$$D = \sqrt{(\cos(A - B) - 1)^2 + (\sin(A - B) - 0)^2}$$

or, $\quad D = \sqrt{\cos^2(A - B) + 1 - 2\cos(A - B) + \sin^2(A - B)}$

Since $\cos^2(A - B) + \sin^2(A - B) = 1$, we get,

$$D = \sqrt{2 - 2\cos(A - B)}$$

or, $\quad D^2 = 2 - 2\cos(A - B)$ $\qquad$ --- (37)

From equations (36) and (37), we get,

$$2 - 2\cos(A - B) = 2 - 2(\cos A \cos B + \sin A \sin B)$$

i.e. $\boxed{\cos(A - B) = (\cos A \cos B + \sin A \sin B)}$ $\qquad$ --- (38)

If we replace the variable B with (-B), we get,

$$\cos\big(A - (-B)\big) = \cos(A + B) = \cos A \cos(-B) + \sin A \sin(-B)$$

Since $\cos(-B) = \cos B$ and $\sin(-B) = -\sin B$, we get,

$$\boxed{\cos(A + B) = \cos A \cos B - \sin A \sin B}$$ $\qquad$ --- (39)

If we put, A = 90° in the equation (38), we get,

$$\cos(90° - B) = \cos 90° \cos B + \sin 90° \sin B$$

Since $\cos 90° = 0$, and $\sin 90° = 1$, we have,

$$\cos(90° - B) = \sin B \qquad\qquad \text{--- (40)}$$

If we replace the variable B with (90°-B) in the equation (40), we get,

$$\cos\big(90° - (90° - B)\big) = \cos B = \sin(90° - B)$$

$$\text{or,} \quad \sin(90° - B) = \cos B \qquad\qquad \text{--- (41)}$$

Now, if we replace the variable B with (A+B) in the equation (40), we get,

$$\cos\big(90° - (A + B)\big) = \cos\big((90° - A) - B\big) = \sin(A + B) \qquad \text{--- (42)}$$

Here, the term $\cos\big((90° - A) - B\big)$ is of the form $\cos(A - B)$. Since,

$$\cos(A - B) = \cos A \cos B + \sin A \sin B$$

We get,

$$\cos\big((90° - A) - B\big) = \cos(90° - A) \cos B + \sin(90° - A) \sin B \qquad \text{--- (43)}$$

From equations (42) and (43), we get,

$$\sin(A + B) = \cos(90° - A) \cos B + \sin(90° - A) \sin B$$

From equations (40) and (41), we have,

$$\cos(90° - A) = \sin A$$

$$\text{and,} \quad \sin(90° - A) = \cos A$$

Therefore, we get,

$$\boxed{\sin(A + B) = \sin A \cos B + \cos A \sin B} \qquad\qquad \text{--- (44)}$$

If we replace the variable B with (-B) in the equation (44), we get,

$$\sin(A - B) = \sin A \cos(-B) + \cos A \sin(-B)$$

Since $\cos(-B) = \cos B$ and $\sin(-B) = -\sin B$, we get,

$$\boxed{\sin(A - B) = \sin A \cos B - \cos A \sin B} \qquad\qquad \text{--- (45)}$$

To summarize, we have,

$$\begin{aligned}
\sin(A + B) &= \sin A \cos B + \cos A \sin B \\
\sin(A - B) &= \sin A \cos B - \cos A \sin B \\
\cos(A + B) &= \cos A \cos B - \sin A \sin B \\
\cos(A - B) &= \cos A \cos B + \sin A \sin B
\end{aligned}$$

These formulae are known as the addition and subtraction formulae. It is important to memorize them thoroughly. If you can remember $\sin(A + B)$, the rest can be easily deduced. To get $\sin(A - B)$ from $\sin(A + B)$, just reverse the sign. i.e., change $(+)$ to $(-)$.

$$\sin(A + B) = \sin A \cos B + \cos A \sin B$$

$$\sin(A - B) = \sin A \cos B - \cos A \sin B$$

To get $\cos(A + B)$ from $\sin(A - B)$, just switch $\sin A$ and $\cos A$. i.e.

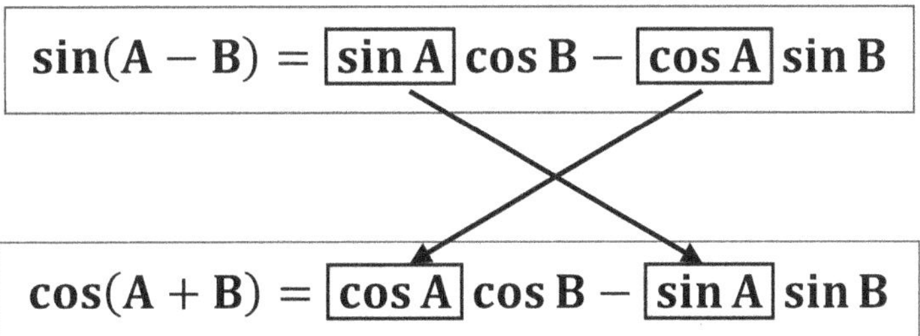

$$\sin(A - B) = \boxed{\sin A}\cos B - \boxed{\cos A}\sin B$$

$$\cos(A + B) = \boxed{\cos A}\cos B - \boxed{\sin A}\sin B$$

And to get $\cos(A - B)$ from $\cos(A + B)$, just reverse the sign. i.e.,

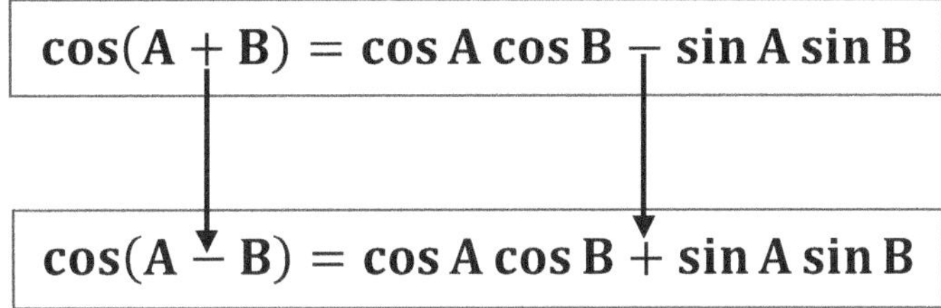

$$\cos(A + B) = \cos A \cos B - \sin A \sin B$$

$$\cos(A - B) = \cos A \cos B + \sin A \sin B$$

$\text{Tan}(A + B)$ can be found from $\sin(A + B)$ and $\cos(A + B)$,

$$\tan(A + B) = \frac{\sin(A + B)}{\cos(A + B)} = \frac{\sin A \cos B + \cos A \sin B}{\cos A \cos B - \sin A \sin B}$$

Dividing both numerator and denominator by $\cos A \cos B$,

$$\tan(A + B) = \frac{\dfrac{\sin A \cos B}{\cos A \cos B} + \dfrac{\cos A \sin B}{\cos A \cos B}}{\dfrac{\cos A \cos B}{\cos A \cos B} - \dfrac{\sin A \sin B}{\cos A \cos B}}$$

$$= \frac{\dfrac{\sin A}{\cos A} + \dfrac{\sin B}{\cos B}}{1 - \dfrac{\sin A \sin B}{\cos A \cos B}}$$

$$\text{i.e.} \quad \boxed{\tan(A + B) = \frac{\tan A + \tan B}{1 - \tan A \tan B}} \qquad \text{--- (46)}$$

$\text{Tan}(A - B)$ can be found from $\sin(A - B)$ and $\cos(A - B)$,

$$\tan(A - B) = \frac{\sin(A - B)}{\cos(A - B)} = \frac{\sin A \cos B - \cos A \sin B}{\cos A \cos B + \sin A \sin B}$$

Again, dividing both numerator and denominator by $\cos A \cos B$,

$$\tan(A - B) = \frac{\dfrac{\sin A \cos B}{\cos A \cos B} - \dfrac{\cos A \sin B}{\cos A \cos B}}{\dfrac{\cos A \cos B}{\cos A \cos B} + \dfrac{\sin A \sin B}{\cos A \cos B}}$$

$$= \frac{\dfrac{\sin A}{\cos A} - \dfrac{\sin B}{\cos B}}{1 + \dfrac{\sin A}{\cos A}\dfrac{\sin B}{\cos B}}$$

$$\text{i.e.} \quad \boxed{\tan(A - B) = \frac{\tan A - \tan B}{1 + \tan A \tan B}} \qquad \text{--- (47)}$$

To summarize, we have,

$$\boxed{\begin{aligned} \tan(A + B) &= \frac{\tan A + \tan B}{1 - \tan A \tan B} \\[2mm] \tan(A - B) &= \frac{\tan A - \tan B}{1 + \tan A \tan B} \end{aligned}}$$

An effective way to improve our memory is by employing a strategy called 'spaced repetition'. The idea is to recall the formulae in repeated sessions with enough time gaps, so that our brain is forced to recognize its importance.

Also, try to recall the previous session before recalling the current, so that our brain recognises where to place each of the formulae in our memory.

Day Five

Session One
Problem session: vol-2

Session Two
Double angle formulae

Session Three
Problem session: vol-3

What you do everyday matters more than what you do once in a while.

- Gretchen Rubin

SESSION ONE

PROBLEM SESSION: VOL-II

Before every practice problem, recall the identities involved, it helps boost your memory. In this session, we shall workout some problems involving the addition and subtraction formulae. So, recall them thoroughly.

Example 12: If $\tan A = \dfrac{3}{4}$ and $\sin B = \dfrac{5}{13}$, where A is in the third and B is in the second quadrant, find

    i. $\sin(A + B)$ ii. $\cos(A + B)$ iii. $\tan(A + B)$

Solution: The first step is to find all the trigonometric functions of both the angles A and B.

Angle A

we have $\tan A = \dfrac{3}{4}$, it's immediate to get

$$\cot A = \dfrac{4}{3}$$

using $1 + \tan^2 A = \sec^2 A$, we get,

$$\sec^2 A = 1 + \dfrac{9}{16} = \dfrac{25}{16}$$

$$\text{i. e. } \sec A = \pm\dfrac{5}{4}$$

since A is in the third quadrant,

$$\sec A = -\dfrac{5}{4}$$

$$\text{therefore, } \cos A = \dfrac{1}{\sec A} = -\dfrac{4}{5}$$

since $\tan A = \dfrac{\sin A}{\cos A}$, we get,

$$\sin A = \tan A \cos A = \left(\dfrac{3}{4}\right)\left(-\dfrac{4}{5}\right) = -\dfrac{3}{5}$$

$$\text{therefore, } \operatorname{cosec} A = \dfrac{1}{\sin A} = -\dfrac{5}{3}$$

To summarize, we have,

$$\sin A = -\frac{3}{5} \qquad \operatorname{cosec} A = -\frac{5}{3}$$

$$\cos A = -\frac{4}{5} \qquad \sec A = -\frac{5}{4}$$

$$\tan A = \frac{3}{4} \qquad \cot A = \frac{4}{3}$$

<u>Angle B</u>

we have $\sin B = \frac{5}{13}$, so $\operatorname{cosec} B = \frac{13}{5}$,

using $\sin^2 B + \cos^2 B = 1$, we get,

$$\cos^2 B = 1 - \frac{25}{169} = \frac{144}{169}$$

$$\text{i. e. } \cos B = \pm\frac{12}{13}$$

since B is in the second quadrant,

$$\cos B = -\frac{12}{13}$$

$$\text{therefore, } \sec B = \frac{1}{\cos B} = -\frac{13}{12}$$

since $\tan B = \dfrac{\sin B}{\cos B}$, we get,

$$\tan B = \frac{\frac{5}{13}}{-\frac{12}{13}} = -\frac{5}{12}$$

$$\cot B = \frac{1}{\tan B} = -\frac{12}{5}$$

To summarize, we have,

$$\sin B = \frac{5}{13} \qquad \operatorname{cosec} B = \frac{13}{5}$$

$$\cos B = -\frac{12}{13} \qquad \sec B = -\frac{13}{12}$$

$$\tan B = -\frac{5}{12} \qquad \cot B = -\frac{12}{5}$$

Now from these values, we get,

i. $\sin(A + B) = \sin A \cos B + \cos A \sin B$

$$= \left(-\frac{3}{5}\right)\left(-\frac{12}{13}\right) + \left(-\frac{4}{5}\right)\left(\frac{5}{13}\right)$$

$$= \frac{36 - 20}{65} = \frac{16}{65}$$

ii. $\cos(A + B) = \cos A \cos B - \sin A \sin B$

$$= \left(-\frac{4}{5}\right)\left(-\frac{12}{13}\right) - \left(-\frac{3}{5}\right)\left(\frac{5}{13}\right)$$

$$= \frac{48 + 15}{65} = \frac{63}{65}$$

iii. $\tan(A + B) = \dfrac{\tan A + \tan B}{1 - \tan A \tan B}$

$$= \frac{\frac{3}{4} + \left(-\frac{5}{12}\right)}{1 - \left(\frac{3}{4}\right)\left(-\frac{5}{12}\right)} = \frac{16}{63}$$

Problem 22:

i. If $\cos A = -\dfrac{12}{13}$ and $\cot B = \dfrac{24}{7}$, where A lies in the second and B lies in the first quadrant. Find: $a)\ \sin(A + B)\,; b)\ \cos(A - B)$.

ii. If $\sin \theta = -\dfrac{3}{5}$ and $\sin \varphi = \dfrac{12}{13}$, where θ lies in the third and φ lies in the second quadrant. Find: $a)\ \tan(\theta - \varphi)\,; b)\ \cot(\theta + \varphi)$.

iii. If $\cos A = \dfrac{3}{5}$ and $\tan B = \dfrac{5}{12}$, where A and B are acute. Find: $a)\ \sin(A + B)\,; b)\ \cos(A - B)$.

Ans:

i. $a)\ \dfrac{36}{325}$; $b)\ -\dfrac{253}{325}$

ii. $a)\ -\dfrac{63}{16}$; $b)\ -\dfrac{56}{33}$

iii. $a)\ \dfrac{63}{65}$; $b)\ \dfrac{56}{65}$

Example 13: Show that $(1 + \tan A)(1 + \tan B) = 2$, if $A + B = 45°$.

Solution: We have, $A + B = 45°$

Therefore,

$$\tan(A + B) = \tan 45° = 1$$

$$\text{i. e.} \quad \frac{\tan A + \tan B}{1 - \tan A \tan B} = 1$$

$$\text{or,} \quad \tan A + \tan B = 1 - \tan A \tan B$$

$$\text{i. e.} \quad \tan A + \tan B + \tan A \tan B = 1$$

Adding 1 on both sides,

$$1 + \tan A + \tan B + \tan A \tan B = 1 + 1$$

$$(1 + \tan A) + \tan B \,(1 + \tan A) = 2$$

Which gives,

$$(1 + \tan A)(1 + \tan B) = 2$$

Problem 23: Show that:

i. $$\frac{\sin(A+B)}{\sin(A-B)} = \frac{\tan A + \tan B}{\tan A - \tan B}$$

[Hint: expand the LHS and divide both numerator and denominator by $\cos A \cos B$.]

ii. $\sin(A + B) \sin(A - B) = \sin^2 A - \sin^2 B$

iii. $\sin\left(\dfrac{\pi}{3} + \theta\right) - \sin\left(\dfrac{\pi}{3} - \theta\right) = \sin\theta$

iv. $\tan 3A \tan 2A \tan A = \tan 3A - \tan 2A - \tan A$

[Hint: start with the equation $\tan 3A = \tan(2A + A)$, and use the $\tan(A + B)$ formula to expand the RHS and do some algebra.]

SESSION TWO

<u>DOUBLE ANGLE FORMULAE</u>

The double angle formulae express $\sin 2\theta$, $\cos 2\theta$ and $\tan 2\theta$ in terms of $\sin\theta$, $\cos\theta$ and $\tan\theta$.

$$\text{i.e.} \quad \mathbf{2\theta} \ \text{ in terms of it's half} - \boldsymbol{\theta}$$

(If you are not interested in the proof, skip it and learn the summary boxes and the equation (54))

We have the addition formula

$$\sin(A + B) = \sin A \cos B + \cos A \sin B$$

Putting $A = B = \theta$ into the formula gives,

$$\sin(\theta + \theta) = \sin\theta\cos\theta + \cos\theta\sin\theta$$

$$\text{or,} \quad \boxed{\sin(2\theta) = 2\sin\theta\cos\theta} \qquad\qquad \text{--- (48)}$$

Here, 2θ is expressed in terms of its half $-$ θ.

Now multiplying and dividing the RHS by $\cos\theta$, we get,

$$\sin(2\theta) = \frac{2\sin\theta\cos\theta}{\cos\theta}\cos\theta$$

$$= 2\tan\theta\cos^2\theta = \frac{2\tan\theta}{\sec^2\theta}$$

Using $1 + \tan^2\theta = \sec^2\theta$, we get,

$$\boxed{\sin(2\theta) = \frac{2\tan\theta}{1 + \tan^2\theta}} \qquad\qquad \text{--- (49)}$$

To summarize,

$$\boxed{\sin(2\theta) = 2\sin\theta\cos\theta = \frac{2\tan\theta}{1 + \tan^2\theta}}$$

Also putting $A = B = \theta$ into the formula

$$\cos(A + B) = \cos A \cos B - \sin A \sin B$$

gives,

$$\cos(2\theta) = \cos^2\theta - \sin^2\theta \qquad\qquad \text{--- (50)}$$

from $\cos^2\theta + \sin^2\theta = 1$ we get, $\sin^2\theta = 1 - \cos^2\theta$. Therefore,

$$\cos(2\theta) = \cos^2\theta - \sin^2\theta = \cos^2\theta - (1 - \cos^2\theta)$$

$$\text{or,} \quad \cos(2\theta) = 2\cos^2\theta - 1 \qquad\qquad \text{--- (51)}$$

Also, from $\cos^2\theta + \sin^2\theta = 1$ we get, $\cos^2\theta = 1 - \sin^2\theta$. Therefore,

$$\cos(2\theta) = \cos^2\theta - \sin^2\theta = 1 - \sin^2\theta - \sin^2\theta$$

$$\text{or,} \quad \cos(2\theta) = 1 - 2\sin^2\theta \qquad\qquad \text{--- (52)}$$

Now,

$$\cos(2\theta) = \cos^2\theta - \sin^2\theta = \cos^2\theta\left(1 - \frac{\sin^2\theta}{\cos^2\theta}\right) = \cos^2\theta\,(1 - \tan^2\theta) = \frac{1 - \tan^2\theta}{\sec^2\theta}$$

Since $1 + \tan^2\theta = \sec^2\theta$, we get,

$$\cos(2\theta) = \frac{1 - \tan^2\theta}{1 + \tan^2\theta} \qquad\qquad \text{--- (53)}$$

To summarize,

$$\boxed{\cos(2\theta) = \cos^2\theta - \sin^2\theta = 2\cos^2\theta - 1 = 1 - 2\sin^2\theta = \frac{1 - \tan^2\theta}{1 + \tan^2\theta}}$$

Also putting $A = B = \theta$ into the equation,

$$\tan(A + B) = \frac{\tan A + \tan B}{1 - \tan A \tan B}$$

gives,

$$\boxed{\tan(2\theta) = \frac{2\tan\theta}{1 - \tan^2\theta}} \qquad\qquad \text{--- (54)}$$

SESSION THREE

PROBLEM SESSION: VOL-III

Problem 24:

i. If $\sin\theta = \dfrac{3}{5}$ and θ is acute. Find: $a)\ \sin 2\theta$; $b)\ \cos 2\theta$; $c)\ \tan 2\theta$

ii. If $\tan A = 1$ and A is in the third quadrant. Find: $a)\ \sin 2A$; $b)\ \cos 2A$; $c)\ \tan 2A$

iii. If $\tan A = \dfrac{3}{4}$ and $\sin B = \dfrac{5}{13}$, where A lies in the first and B lies in the second quadrant. Find: $a)\ \sin 2A$; $b)\ \sin 2B$; $c)\ \cos 2A$

Ans:

i. $a)\ \dfrac{24}{25}$; $b)\ \dfrac{7}{25}$; $c)\ \dfrac{24}{7}$

ii. $a)\ 1$; $b)\ 0$; $c)$ Not Defined

iii. $a)\ \dfrac{24}{25}$; $b)\ -\dfrac{120}{169}$; $c)\ \dfrac{7}{25}$

Example 14: Prove that

$$\frac{\cot\theta - \tan\theta}{\cot\theta + \tan\theta} = \cos 2\theta$$

Solution: We have,

$$\text{LHS} = \frac{\cot\theta - \tan\theta}{\cot\theta + \tan\theta} = \frac{\dfrac{\cos\theta}{\sin\theta} - \dfrac{\sin\theta}{\cos\theta}}{\dfrac{\cos\theta}{\sin\theta} + \dfrac{\sin\theta}{\cos\theta}} = \frac{\dfrac{\cos^2\theta - \sin^2\theta}{\cos\theta\sin\theta}}{\dfrac{\cos^2\theta + \sin^2\theta}{\cos\theta\sin\theta}} = \frac{\cos^2\theta - \sin^2\theta}{\cos^2\theta + \sin^2\theta}$$

Since, $\cos^2\theta + \sin^2\theta = 1$ and $\cos^2\theta - \sin^2\theta = \cos 2\theta$, we get

$$\text{LHS} = \cos 2\theta = \text{RHS}$$

<u>**Problem 25**</u>: **Prove the following identities.**

i. $\dfrac{\sin 2\theta}{1+\cos 2\theta} = \tan\theta$

ii. $\cos^4\theta - \sin^4\theta = \cos 2\theta$

iii. $\dfrac{1-\cos 2\theta}{2\sin\theta} = \sin\theta$

iv. $1 + \tan\theta\tan 2\theta = \sec 2\theta$

v. $\dfrac{\sin 2\theta}{1-\cos 2\theta} = \cot\theta$

vi. $(\sin\theta + \cos\theta)^2 = 1 + \sin 2\theta$

vii. $\dfrac{1-\cos 2\theta}{1+\cos 2\theta} = \tan^2\theta$

Day Six

Session One

Half angle formulae

Triple angle formulae

Session Two

Sum to product formulae: set-1

Session Three

Sum to product formulae: set-2

Do not worry about your difficulties in mathematics, mine are still greater.

- Albert Einstein

SESSION ONE

HALF ANGLE FORMULAE

Multiple angle formulae express $\sin 2\theta, \cos 2\theta$ and $\tan 2\theta$ in terms of $\sin \theta, \cos \theta$ and $\tan \theta$.

$$\text{i.e.} \quad \mathbf{2\theta} \text{ in terms of it's half} - \mathbf{\theta}$$

The half angle formulae express $\sin \theta, \cos \theta$ and $\tan \theta$ in terms of $\sin \left(\frac{\theta}{2}\right), \cos \left(\frac{\theta}{2}\right)$ and $\tan \left(\frac{\theta}{2}\right)$.

$$\text{i.e.} \quad \mathbf{\theta} \text{ in terms of it's half} - \mathbf{\frac{\theta}{2}}$$

(If you are not interested in the proof, then skip it and learn the key-point and the comparison box)

We have the double angle formula.

$$\sin(2\theta) = 2 \sin \theta \cos \theta = \frac{2 \tan \theta}{1 + \tan^2 \theta}$$

Put $2\theta = A$, i.e., $\theta = \dfrac{A}{2}$

$$\sin(A) = 2 \sin \left(\frac{A}{2}\right) \cos \left(\frac{A}{2}\right) = \frac{2 \tan \left(\frac{A}{2}\right)}{1 + \tan^2 \left(\frac{A}{2}\right)}$$

For uniformity with the multiple angle formula, we shall replace A by θ, to obtain the half angle formula,

$$\boxed{\ \sin(\theta) = 2 \sin \left(\frac{\theta}{2}\right) \cos \left(\frac{\theta}{2}\right) = \frac{2 \tan \left(\frac{\theta}{2}\right)}{1 + \tan^2 \left(\frac{\theta}{2}\right)}\ } \qquad \text{--- (55)}$$

By comparison with the double angle formulae, we observe that the half angle formula can be obtained by replacing every angle by its half. That is,

$$\boxed{\begin{aligned} \cos(\theta) &= \cos^2 \left(\frac{\theta}{2}\right) - \sin^2 \left(\frac{\theta}{2}\right) = 2 \cos^2 \left(\frac{\theta}{2}\right) - 1 \\[2mm] &= 1 - 2 \sin^2 \left(\frac{\theta}{2}\right) = \frac{1 - \tan^2 \left(\frac{\theta}{2}\right)}{1 + \tan^2 \left(\frac{\theta}{2}\right)} \end{aligned}} \qquad \text{--- (56)}$$

and,

$$tan(\theta) = \frac{2 \tan\left(\frac{\theta}{2}\right)}{1 - \tan^2\left(\frac{\theta}{2}\right)}$$ --- (57)

> **Key point:** To get the half angle formulae from the double angle formulae, replace θ by its half : $\dfrac{\theta}{2}$

COMPARISON BETWEEN DOUBLE AND HALF ANGLE FORMULAE

DOUBLE ANGLE	HALF ANGLE
$\sin(2\theta) = 2 \sin\theta \cos\theta = \dfrac{2 \tan\theta}{1 + \tan^2\theta}$	$\sin(\theta) = 2 \sin\left(\frac{\theta}{2}\right) \cos\left(\frac{\theta}{2}\right) = \dfrac{2 \tan\left(\frac{\theta}{2}\right)}{1 + \tan^2\left(\frac{\theta}{2}\right)}$
$\begin{aligned} \cos(2\theta) &= \cos^2\theta - \sin^2\theta \\ &= 2\cos^2\theta - 1 \\ &= 1 - 2\sin^2\theta \\ &= \frac{1 - \tan^2\theta}{1 + \tan^2\theta} \end{aligned}$	$\begin{aligned} \cos(\theta) &= \cos^2\left(\frac{\theta}{2}\right) - \sin^2\left(\frac{\theta}{2}\right) \\ &= 2\cos^2\left(\frac{\theta}{2}\right) - 1 \\ &= 1 - 2\sin^2\left(\frac{\theta}{2}\right) \\ &= \frac{1 - \tan^2\left(\frac{\theta}{2}\right)}{1 + \tan^2\left(\frac{\theta}{2}\right)} \end{aligned}$
$\tan(2\theta) = \dfrac{2 \tan\theta}{1 - \tan^2\theta}$	$\tan(\theta) = \dfrac{2 \tan\left(\frac{\theta}{2}\right)}{1 - \tan^2\left(\frac{\theta}{2}\right)}$

Example 15: Find the value of cos 120° using half angle formulae.

Solution: We have,

$$\cos(\theta) = 2\cos^2\left(\frac{\theta}{2}\right) - 1$$

Putting $\theta = 120°$ gives,

$$\cos(120°) = 2\cos^2\left(\frac{120°}{2}\right) - 1 = 2\cos^2(60°) - 1$$

$$= 2\left(\frac{1}{2}\right)^2 - 1 = 2\left(\frac{1}{4}\right) - 1 = \frac{1}{2} - 1 = -\frac{1}{2}$$

Problem 26: Find the value of the following trigonometric functions:

    i.   cos 15°   ii.   cos 135°

Ans:

    i. $\sqrt{\frac{1}{2}\left(\frac{\sqrt{3}}{2} + 1\right)}$ ii. $-\frac{1}{\sqrt{2}}$

TRIPLE ANGLE FORMULAE

Triple angle formulae express $\sin(3\theta)$ and $\cos(3\theta)$ in terms of $\sin(\theta)$ and $\cos(\theta)$.

(If you are not interested in the proof, skip it and learn the equations (58) and (59))

We can write $\sin(3\theta)$ as $\sin(2\theta + \theta)$, and use the sum formula to get,

$$\sin(3\theta) = \sin(2\theta + \theta)$$

$$= \sin(2\theta)\cos\theta + \cos(2\theta)\sin\theta$$

Using the double angle formula, we get,

$$\sin(3\theta) = (2\sin\theta\cos\theta)\cos\theta + (1 - 2\sin^2\theta)\sin\theta$$

$$= 2\sin\theta\cos^2\theta + \sin\theta - 2\sin^3\theta$$

Using, $\cos^2 \theta = 1 - \sin^2 \theta$, we get,

$$\sin(3\theta) = 2 \sin \theta - 2 \sin^3 \theta + \sin \theta - 2 \sin^3 \theta$$

$$\text{or,} \quad \boxed{\sin(3\theta) = 3 \sin \theta - 4 \sin^3 \theta} \qquad \text{--- (58)}$$

Also, we can write $\cos(3\theta)$ as $\cos(2\theta + \theta)$ to get,

$$\cos(3\theta) = \cos(2\theta) \cos \theta - \sin(2\theta) \sin \theta$$

$$= (2 \cos^2 \theta - 1) \cos \theta - (2 \sin \theta \cos \theta) \sin \theta$$

$$= 2 \cos^3 \theta - \cos \theta - 2 \cos \theta \, (1 - \cos^2 \theta)$$

$$\text{or,} \quad \boxed{\cos(3\theta) = 4 \cos^3 \theta - 3 \cos \theta} \qquad \text{--- (59)}$$

To summarize,

$$\sin(3\theta) = 3 \sin \theta - 4 \sin^3 \theta$$

$$\cos(3\theta) = 4 \cos^3 \theta - 3 \cos \theta$$

To get $\cos(3\theta)$ from $\sin(3\theta)$, replace sine by cosine and switch their positions. i.e.,

$$\mathbf{\sin(3\theta) = 3 \, \boxed{\sin \theta} - 4 \, \boxed{\sin^3 \theta}}$$

$$\mathbf{\cos(3\theta) = 4 \, \boxed{\cos^3 \theta} - 3 \, \boxed{\cos \theta}}$$

<u>Example 16</u>: **Prove that,**

$$\frac{\cos(3\theta) + \cos \theta}{\sin(3\theta) - \sin \theta} = \cot \theta$$

<u>Solution</u>: We have,

$$\cos(3\theta) + \cos \theta = 4 \cos^3 \theta - 3 \cos \theta + \cos \theta$$
$$= 2 \cos \theta \, (2 \cos^2 \theta - 1) = 2 \cos \theta \cos(2\theta)$$

and,

$$\sin(3\theta) - \sin\theta = 3\sin\theta - 4\sin^3\theta - \sin\theta$$
$$= 2\sin\theta\,(1 - 2\sin^2\theta) = 2\sin\theta\cos(2\theta)$$

Therefore,

$$\frac{\cos(3\theta) + \cos\theta}{\sin(3\theta) - \sin\theta} = \frac{2\cos\theta\cos(2\theta)}{2\sin\theta\cos(2\theta)} = \cot\theta$$

Problem 27: Prove the following identities.

i. $\dfrac{\sin 3\theta + \sin\theta}{\cos 3\theta - \cos\theta} = -\cot\theta$
ii. $\dfrac{\sin 3\theta}{\sin\theta} + \dfrac{\cos 3\theta}{\cos\theta} = 4\cos 2\theta$

iii. $\dfrac{\sin 3\theta}{\sin\theta} - \dfrac{\cos 3\theta}{\cos\theta} = 2$
iv. $\tan 3\theta = \dfrac{3\tan\theta - \tan^3\theta}{1 - 3\tan^2\theta}$

SESSION TWO

SUM TO PRODUCT FORMULAE: SET-1

We have the addition and subtraction formulae,

$$\sin(A + B) = \sin A\cos B + \cos A\sin B \qquad\qquad \text{--- (44)}$$

$$\sin(A - B) = \sin A\cos B - \cos A\sin B \qquad\qquad \text{--- (45)}$$

Adding equations (44) and (45) gives,

$$\boxed{\sin(A + B) + \sin(A - B) = 2\sin A\cos B} \qquad\qquad \text{--- (60)}$$

This equation connects a sum to a product, hence the name: 'Sum to Product'.

Also, subtracting equation (45) and (44) gives,

$$\boxed{\sin(A + B) - \sin(A - B) = 2\cos A\sin B} \qquad\qquad \text{--- (61)}$$

Similarly, from the addition and subtraction formulae for $\cos(A + B)$ and $\cos(A - B)$, we get the equations,

$$\boxed{\cos(A + B) + \cos(A - B) = 2\cos A \cos B} \qquad \text{--- (62)}$$

$$\boxed{\cos(A + B) - \cos(A - B) = -2\sin A \sin B} \qquad \text{--- (63)}$$

To summarize, we have,

$$
\begin{aligned}
\sin(A + B) + \sin(A - B) &= 2\sin A \cos B \\
\sin(A + B) - \sin(A - B) &= 2\cos A \sin B \\
\cos(A + B) + \cos(A - B) &= 2\cos A \cos B \\
\cos(A + B) - \cos(A - B) &= -2\sin A \sin B
\end{aligned}
$$

To memorize these equations, we shall group them into two categories,

- The sine category
- The cosine category

Consider the sine category,

$$\sin(A + B) + \sin(A - B) = 2\sin A \cos B$$

$$\sin(A + B) - \sin(A - B) = 2\cos A \sin B$$

The LHS of these equations contain two terms: $\sin(A + B)$ and $\sin(A - B)$. The only distinction is that in the first equation, we add them. That is,

$$\sin(A + B) + \sin(A - B)$$

and in the second, we subtract them. That is,

$$\sin(A + B) - \sin(A - B)$$

The RHS of these equations are the two terms in the RHS of the addition formula,

$$\sin(A + B) = \sin A \cos B + \cos A \sin B \text{ times 2.}$$

That is,

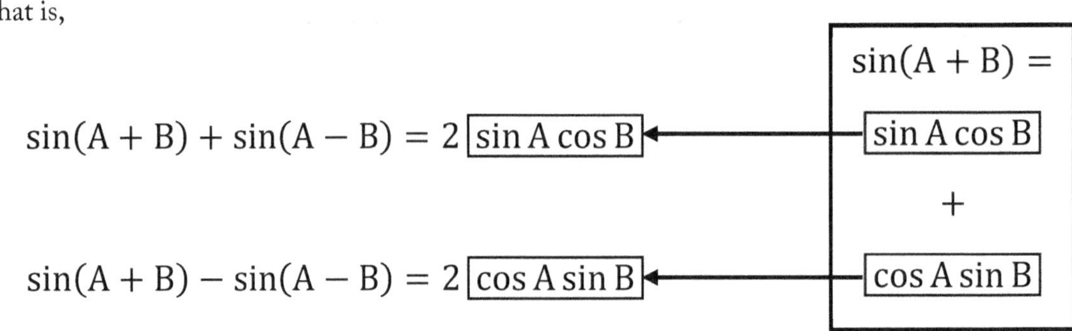

Now, consider the cosine category,

$$\cos(A + B) + \cos(A - B) = 2 \cos A \cos B$$

$$\cos(A + B) - \cos(A - B) = -2 \sin A \sin B$$

The LHS of these equations contain only two terms, $\cos(A + B)$ and $\cos(A - B)$. Similar to the sine category, in the first equation, we add them. That is,

$$\cos(A + B) + \cos(A - B)$$

And in the second, we subtract them. that is,

$$\cos(A + B) - \cos(A - B)$$

The RHS of these equations are the two terms in the RHS of the addition formula

$$\cos(A + B) = \cos A \cos B - \sin A \sin B \text{ times 2.}$$

That is,

$$\cos(A + B) + \cos(A - B) = 2 \boxed{\cos A \cos B}$$

$$\cos(A + B) - \cos(A - B) = -2 \boxed{\sin A \sin B}$$

$$\cos(A + B) = \boxed{\cos A \cos B} + \boxed{- \sin A \sin B}$$

SESSION THREE

<u>SUM TO PRODUCT FORMULAE: SET-II</u>

In this session, we shall rewrite the sum to product formulae in a different form.

We have,

$$\sin(A + B) + \sin(A - B) = 2 \sin A \cos B$$

Put,

$$A + B = \theta \qquad \qquad \text{--- (64)}$$

$$A - B = \varphi \qquad \qquad \text{--- (65)}$$

Such that, $(64) + (65)$ gives,

$$A \quad + \quad B \quad = \quad \theta$$
$$+ \quad A \quad - \quad B \quad = \quad \varphi$$
$$\overline{}$$
$$2A \quad = \quad \theta \quad + \quad \varphi$$

or,

$$A = \frac{\theta + \varphi}{2}$$

And $(64) - (65)$ gives,

$$A \quad + \quad B \quad = \quad \theta$$
$$- \quad (A \quad - \quad B \quad = \quad \varphi)$$
$$\overline{}$$
$$2B \quad = \quad \theta \quad - \quad \varphi$$

or,

$$B = \frac{\theta - \varphi}{2}$$

Therefore, the sum to product formula becomes,

$$\sin \theta + \sin \varphi = 2 \sin \left(\frac{\theta + \varphi}{2} \right) \cos \left(\frac{\theta - \varphi}{2} \right)$$

For uniformity with the sum to product formula, we shall write,

$$\boxed{\sin A + \sin B = 2 \sin \left(\frac{A + B}{2} \right) \cos \left(\frac{A - B}{2} \right)}$$

Similarly, we get,

$$\boxed{\begin{aligned}
\sin A - \sin B &= 2 \cos \left(\frac{A + B}{2} \right) \sin \left(\frac{A - B}{2} \right) \\
\cos A + \cos B &= 2 \cos \left(\frac{A + B}{2} \right) \cos \left(\frac{A - B}{2} \right) \\
\cos A - \cos B &= -2 \sin \left(\frac{A + B}{2} \right) \sin \left(\frac{A - B}{2} \right)
\end{aligned}}$$

These formulae can be memorized by methods similar to that of the sum to product formulae: set-I

<table>
<tr><td colspan="2" align="center">COMPARISON BETWEEN PRODUCT FORMULAE

SET-I AND SET-II</td></tr>
<tr><td align="center">SET-I</td><td align="center">SET-II</td></tr>
<tr>
<td>

$\sin(A + B) + \sin(A - B) = 2\sin A \cos B$

$\sin(A + B) - \sin(A - B) = 2\cos A \sin B$

$\cos(A + B) + \cos(A - B) = 2\cos A \cos B$

$\cos(A + B) - \cos(A - B) = -2\sin A \sin B$

</td>
<td>

$\sin A + \sin B = 2\sin\left(\dfrac{A + B}{2}\right)\cos\left(\dfrac{A - B}{2}\right)$

$\sin A - \sin B = 2\cos\left(\dfrac{A + B}{2}\right)\sin\left(\dfrac{A - B}{2}\right)$

$\cos A + \cos B = 2\cos\left(\dfrac{A + B}{2}\right)\cos\left(\dfrac{A - B}{2}\right)$

$\cos A - \cos B = -2\sin\left(\dfrac{A + B}{2}\right)\sin\left(\dfrac{A - B}{2}\right)$

</td>
</tr>
</table>

Day Seven

Session One

Problem session: vol-4

Session Two

Allied angles

Session Three

Problem session: vol-5

Every task, goal, race and year comes to an end. Therefore, make a habit to always finish strong.

- Gary Ryan Blair

SESSION ONE

PROBLEM SESSION: VOL-IV

In this session, we shall workout some problems involving the sum to product formulae. So, recall them thoroughly.

Example 17: Express $\cos 50° \cos 70°$ as a sum.

Solution: We have,

$$\cos(A + B) + \cos(A - B) = 2 \cos A \cos B$$

or,

$$\cos A \cos B = \frac{1}{2}\left(\cos(A + B) + \cos(A - B)\right)$$

therefore,

$$\cos 50° \cos 70° = \frac{1}{2}\left(\cos(50° + 70°) + \cos(50° - 70°)\right)$$

$$= \frac{1}{2}\left(\cos(120°) + \cos(-20°)\right)$$

Since, $\cos(-x) = \cos x$, we get $\cos(-20°) = \cos 20°$. Therefore,

$$\cos 50° \cos 70° = \frac{1}{2}\left(\cos(120°) + \cos(20°)\right)$$

Thus, the product $\cos 50° \cos 70°$ is expressed as the sum

$$\frac{1}{2}\left(\cos(120°) + \cos(20°)\right)$$

Problem 28: Express the following products as a sum or a difference:

    i. $\cos 25° \cos 75°$ ii. $\sin 35° \cos 15°$ iii. $\sin 40° \sin 20°$

    iv. $\sin 5\theta \cos 3\theta$ v. $\cos 3A \cos A$ vi. $\cos 7\theta \cos 4\theta$

Example 18: Express $\sin 50° + \sin 20°$ as a product.

Solution: We have,

$$\sin A + \sin B = 2 \sin \left(\frac{A + B}{2}\right) \cos \left(\frac{A - B}{2}\right)$$

therefore,

$$\sin 50° + \sin 20° = 2 \sin \left(\frac{50° + 20°}{2}\right) \cos \left(\frac{50° - 20°}{2}\right)$$
$$= 2 \sin 35° \cos 15°$$

Thus, the sum $\sin 50° + \sin 20°$ is expressed as a product $2 \sin 35° \cos 15°$.

Problem 29: Express the following as products:

   i. $\cos 15° + \cos 65°$ ii. $\sin 75° - \sin 35°$ iii. $\sin 60° + \sin 40°$

   iv. $\sin 4\theta + \sin 2\theta$ v. $\sin 5\theta - \sin 3\theta$ vi. $\cos 8\theta - \cos 4\theta$

Example 19: Prove the following equation:

$$\cos 80° + \cos 40° - \cos 20° = 0$$

Solution: Applying the sum to product formula only to the first two terms of the LHS gives,

$$\text{LHS} = 2 \cos \left(\frac{80° + 40°}{2}\right) \cos \left(\frac{80° - 40°}{2}\right) - \cos 20°$$
$$= 2 \cos 60° \cos 20° - \cos 20°$$

Since, $\cos 60° = \dfrac{1}{2}$ we get,

$$\text{LHS} = 2 \left(\frac{1}{2}\right) \cos 20° - \cos 20°$$
$$= \cos 20° - \cos 20° = 0 = \text{RHS}$$

Problem 30: Prove the following equations:

i. $\cos 21° - \cos 7° = -2 \sin 14° \sin 7°$

ii. $\sin 35° + \sin 25° = \cos 5°$

iii. $\sin 50° - \sin 70° + \sin 10° = 0$

iv. $\sin 33° + \cos 63° = \cos 3°$

v. $\sin 50° - \sin 70° + \cos 80° = 0$

vi. $\cos\left(\dfrac{\pi}{8}\right) + \cos\left(\dfrac{3\pi}{8}\right) + \cos\left(\dfrac{5\pi}{8}\right) + \cos\left(\dfrac{7\pi}{8}\right) = 0$

vii. $\sin\theta + \sin 3\theta + \sin 5\theta + \sin 7\theta = 4\cos\theta\cos 2\theta\sin 4\theta$

Example 20: Prove

$$\frac{\cos 3\theta + \cos\theta}{\sin 3\theta - \sin\theta} = \cot\theta$$

using sum to product formulae.

Solution:

$$\text{LHS} = \frac{\cos 3\theta + \cos\theta}{\sin 3\theta - \sin\theta} = \frac{2\cos\left(\dfrac{3\theta + \theta}{2}\right)\cos\left(\dfrac{3\theta - \theta}{2}\right)}{2\cos\left(\dfrac{3\theta + \theta}{2}\right)\sin\left(\dfrac{3\theta - \theta}{2}\right)}$$

$$= \frac{2\cos 2\theta\cos\theta}{2\cos 2\theta\sin\theta} = \frac{\cos\theta}{\sin\theta} = \cot\theta = \text{RHS}$$

Problem 31: Prove the following identities:

i. $\dfrac{\sin 3\theta + \sin\theta}{\cos 3\theta - \cos\theta} = -\cot\theta$

ii. $\dfrac{\sin 5\theta + \sin 3\theta}{\cos 5\theta + \cos 3\theta} = \tan 4\theta$

iii. $\dfrac{\sin\theta + \sin 2\theta}{\cos\theta + \cos 2\theta} = \tan\left(\dfrac{3\theta}{2}\right)$

iv. $\dfrac{\cos 7\theta + \cos 5\theta}{\sin 7\theta - \sin 5\theta} = \cot\theta$

Example 21: Prove

$$\frac{\sin 2\theta + \sin 5\theta - \sin \theta}{\cos 2\theta + \cos 5\theta + \cos \theta} = \tan 2\theta$$

Solution:

$$\text{LHS} = \frac{\sin 2\theta + \sin 5\theta - \sin \theta}{\cos 2\theta + \cos 5\theta + \cos \theta} = \frac{\sin 2\theta + 2 \cos \left(\dfrac{5\theta + \theta}{2}\right) \sin \left(\dfrac{5\theta - \theta}{2}\right)}{\cos 2\theta + 2 \cos \left(\dfrac{5\theta + \theta}{2}\right) \cos \left(\dfrac{5\theta - \theta}{2}\right)}$$

$$= \frac{\sin 2\theta + 2 \cos 3\theta \sin 2\theta}{\cos 2\theta + 2 \cos 3\theta \cos 2\theta} = \frac{\sin 2\theta \, (1 + 2 \cos 3\theta)}{\cos 2\theta \, (1 + 2 \cos 3\theta)}$$

$$= \frac{\sin 2\theta}{\cos 2\theta} = \tan 2\theta = \text{RHS}$$

Problem 32: Prove the following identities:

i. $\dfrac{\cos 4\theta + \cos 3\theta + \cos 2\theta}{\sin 4\theta + \sin 3\theta + \sin 2\theta} = \cot 3\theta$ ii. $\dfrac{\sin \theta + \sin 3\theta + \sin 5\theta}{\cos \theta + \cos 3\theta + \cos 5\theta} = \tan 3\theta$

iii. $\dfrac{\sin 5\theta - 2 \sin 3\theta + \sin \theta}{\cos 5\theta - \cos \theta} = \tan \theta$

SESSION TWO

ALLIED ANGLES

In this session we shall look at angles that can be expressed as the sum of a quadrant angle (0°, 90°, 180°, 270° and 360°) and a standard acute angle (30°, 45° and 60°).

For instance, the angle 150° can be expressed as,

$$150° = 90° + 60°$$

where, 90° is the quadrant angle and 60° is the standard acute angle (see figure 48).

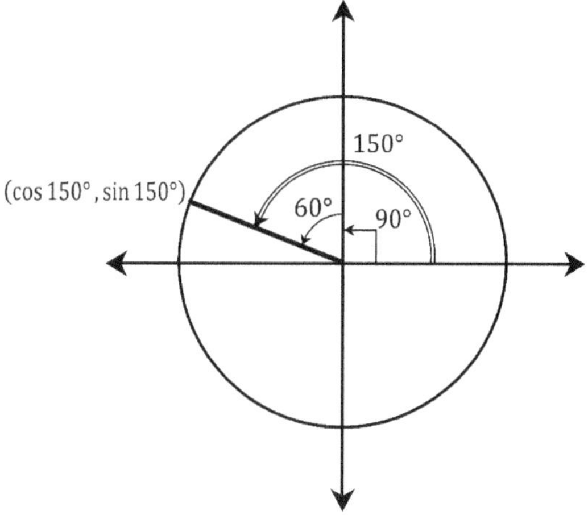

Figure 48

Similarly, the angle 240° can be expressed as,

$$240° = 180° + 60°$$

where, 180° is the quadrant angle and 60° is the standard acute angle (see figure 49).

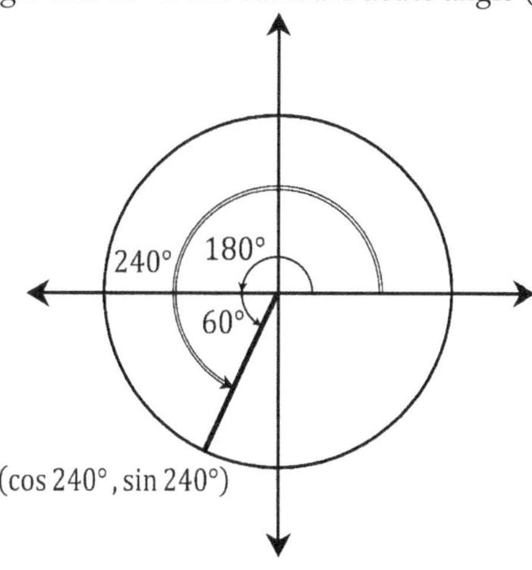

Figure 49

The angle 240° can also be expressed as,

$$240° = 270° - 30°$$

where, 270° is the quadrant angle and 30° is the standard acute angle (see figure 50).

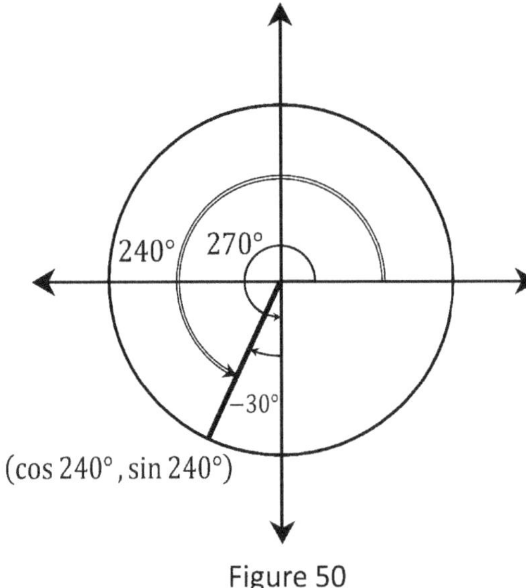

Figure 50

Such angles are called Allied angles.

Consider such an allied angle in the first quadrant,

$$90° - \theta$$

Here, 90° is the quadrant angle and θ is the acute angle (see figure 51).

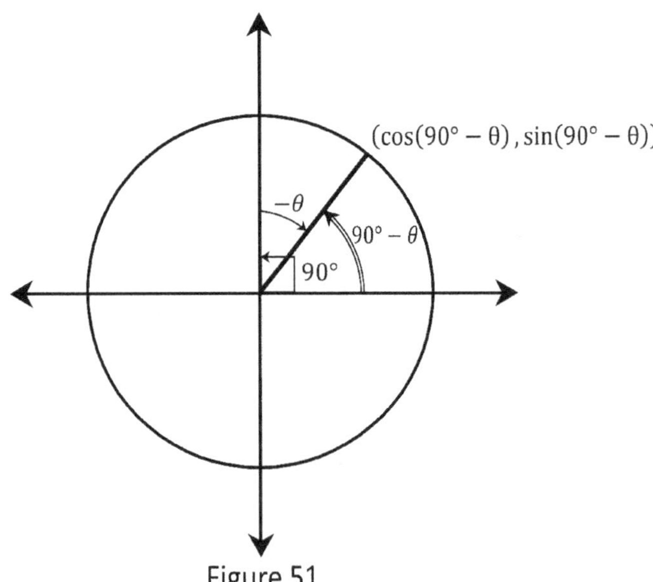

Figure 51

Using the subtraction formula, we get,

$$\sin(90° - \theta) = \sin 90° \cos \theta - \cos 90° \sin \theta$$

Since, $\sin 90° = 1$ and $\cos 90° = 0$, we have,

$$\boxed{\sin(90° - \theta) = \cos \theta}$$

Similarly,

$$\cos(90° - \theta) = \cos 90° \cos \theta + \sin 90° \sin \theta$$

or,

$$\boxed{\cos(90° - \theta) = \sin \theta}$$

To summarize,

$$\sin(90° - \theta) = \cos \theta$$
$$\cos(90° - \theta) = \sin \theta$$

That is, whenever an acute angle is subtracted from $90°$ sine flips to cosine and cosine flips to sine.

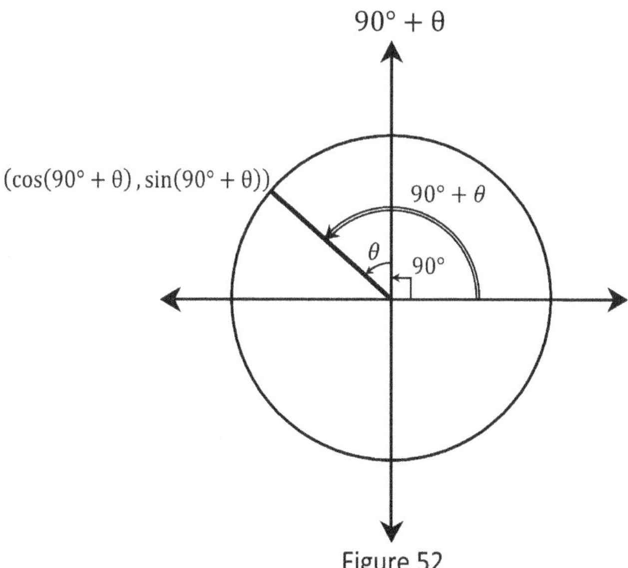

Now consider an allied angle in the second quadrant (see figure 52),

Figure 52

We have,

$$\sin(90° + \theta) = \sin 90° \cos \theta + \cos 90° \sin \theta = \cos \theta$$

$$\cos(90° + \theta) = \cos 90° \cos \theta - \sin 90° \sin \theta = -\sin \theta$$

To summarize,

$$\sin(90° + \theta) = \cos \theta$$
$$\cos(90° + \theta) = -\sin \theta$$

That is, whenever an acute angle (θ) is added or subtracted with 90°, sine flips to cosine and cosine flips to sine.

But there is a distinction between $90° + \theta$ and $90° - \theta$. There is a negative sign in front of $\sin \theta$ corresponding to $\cos(90° + \theta)$.

$90° - \theta$	**$90° + \theta$**
$\sin(90° - \theta) = \cos \theta$ $\cos(90° - \theta) = \sin \theta$	$\sin(90° + \theta) = \cos \theta$ $\cos(90° + \theta) = -\sin \theta$

This is because of the fact that cosine is negative in the second quadrant.

Using similar methods, the allied angles corresponding to the quadrant angles, 180°, 270° and 360° can also be found.

TABLE OF ALLIED ANGLES	
$\sin(90° - \theta) = \cos \theta$ $\cos(90° - \theta) = \sin \theta$	$\sin(90° + \theta) = \cos \theta$ $\cos(90° + \theta) = -\sin \theta$
$\sin(180° - \theta) = \sin \theta$ $\cos(180° - \theta) = -\cos \theta$	$\sin(180° + \theta) = -\sin \theta$ $\cos(180° + \theta) = -\cos \theta$
$\sin(270° - \theta) = -\cos \theta$ $\cos(270° - \theta) = -\sin \theta$	$\sin(270° + \theta) = -\cos \theta$ $\cos(270° + \theta) = \sin \theta$
$\sin(360° - \theta) = -\sin \theta$ $\cos(360° - \theta) = \cos \theta$	$\sin(360° + \theta) = \sin \theta$ $\cos(360° + \theta) = \cos \theta$

Even though the table seem big, there is a simple shortcut.

For instance, consider the angle,

$$270° + \theta$$

Here, the quadrant angle is 270°.

The first step is to find the axis corresponding to the quadrant angle.

> **If the angle is in the X-axis, sine and cosine remains the same.**
>
> **If the angle is in the Y-axis, sine and cosine flip each other.**

We know that 270° is in the Y-axis.

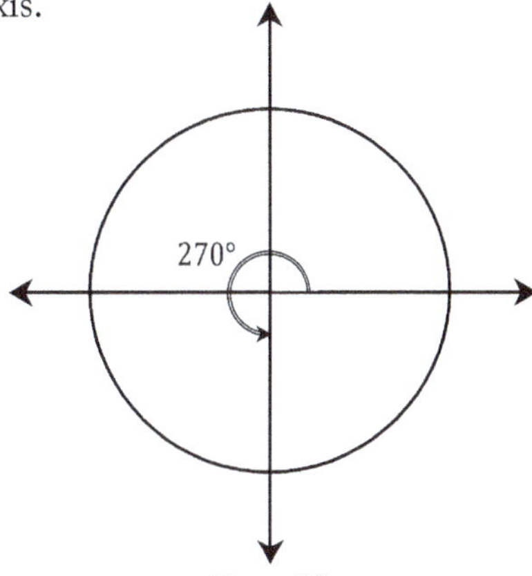

Figure 53

Therefore, sine and cosine flip each other. i.e.,

$$\sin(270° + \theta) = (?)\cos\theta$$

$$\cos(270° + \theta) = (?)\sin\theta$$

The next step is to find the sign in front of the functions on the RHS. For that, two things must be considered.

    a) The function on the LHS.

    b) The quadrant in which the allied angle lies.

Then,

> **The sign of the function on RHS is the sign of the function on the LHS in the quadrant in which the allied angle lies.**

In the equation,

$$\sin(270° + \theta) = (?)\cos\theta$$

The function on the LHS is sine,

$$\boxed{\sin}(270° + θ) = (?)\cos θ$$

and the quadrant in which the allied angle $(270° + θ)$ lies is IV

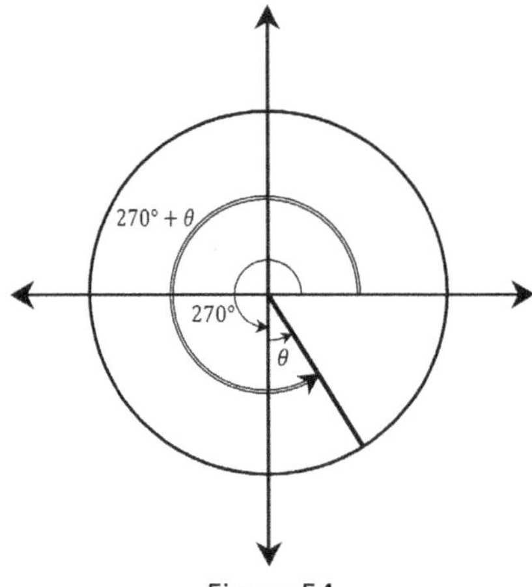

Figure 54

Since the sign of sine is the IV quadrant is negative, we have,

$$\sin(270° + θ) = -\cos θ$$

Similarly, in the equation,

$$\cos(270° + θ) = (?)\sin θ$$

the function on the LHS is cosine and the quadrant in which the allied angle lies is IV. Since the sign of cosine in the IV quadrant is positive, we get,

$$\cos(270° + θ) = +\sin θ$$

Let's look at another example. Consider the function,

$$\sin(180° + θ)$$

Where, the quadrant angle is 180°, which is in the X-axis. Therefore, there is no flip. That is,

$$\sin(180° + θ) = (?)\sin θ$$

Here the function on the LHS is sine and the allied angle $(180° + θ)$ lies in the III quadrant. Since sine is negative in the third quadrant, we get,

$$\sin(180° + θ) = -\sin θ$$

Tangent of an allied angle can be expressed as the ratio of sine and cosine. For instance,

$$\tan(270° + θ) = \frac{\sin(270° + θ)}{\cos(270° + θ)} = \frac{-\cos θ}{\sin θ} = -\cot θ$$

Problem 33: Using the shortcut method, find the functions in the table of allied angles. Also find their tangents.

SESSION THREE

PROBLEM SESSION: VOL-V

In this session, we shall workout some problems involving allied angles.

Example 22: Find the value of $\cos 120°$ using the allied angle method.

Solution: We can write $\cos 120°$ as $\cos(90° + 30°)$.

Since $90°$ is in the Y- axis, cosine flips into sine.

$$\cos(90° + 30°) = (?) \sin 30°$$

And since, $120°$ is in the second quadrant, where cosine is negative, we get,

$$\cos(90° + 30°) = - \sin 30° = -\frac{1}{2}.$$

Since, $120°$ can also be written as $180° - 60°$. We can also write,

$$\cos 120° = \cos(180° - 60°) = (?) \cos 60°$$

And since $120°$ is in the second quadrant, where cosine is negative, we get,

$$\cos 120° = \cos(180° - 60°) = - \cos 60° = -\frac{1}{2}$$

Problem 34: Find the values of:

    i. $\sin 120°$ ii. $\tan 150°$ iii. $\tan 135°$ iv. $\sin 150°$

    v. $\sin 300°$ vi. $\sin 210°$ vii. $\cos 240°$ viii. $\tan 330°$

Ans:

    i. $\dfrac{\sqrt{3}}{2}$ ii. $-\dfrac{1}{\sqrt{3}}$ iii. -1 iv. $\dfrac{1}{2}$

    v. $-\dfrac{\sqrt{3}}{2}$ vi. $-\dfrac{1}{2}$ vii. $-\dfrac{1}{2}$ viii. $-\dfrac{1}{\sqrt{3}}$

Example 23: Find the value of $\cos 750°$ using the allied angle method.

Solution: We can write $\cos 750°$ as $\cos(360° + 360° + 30°)$.

We have $\cos(360° + \theta) = \cos\theta$

Therefore,

$$\cos 750° = \cos\big(360° + (360° + 30°)\big) = \cos(360° + 30°)$$

Again using $\cos(360° + \theta) = \cos\theta$, we get,

$$\cos(360° + 30°) = \cos(30°) = \frac{\sqrt{3}}{2}$$

That is,

$$\cos 750° = \frac{\sqrt{3}}{2}$$

Problem 35: Find the values of:

   i. $\cos 390°$ ii. $\sin 330°$ iii. $\cos 570°$ iv. $\sin 510°$

   v. $\sin 1200°$

Ans:

   i. $\dfrac{\sqrt{3}}{2}$ ii. $-\dfrac{1}{2}$ iii. $-\dfrac{\sqrt{3}}{2}$ iv. $\dfrac{1}{2}$ v. $\dfrac{\sqrt{3}}{2}$

The scope and applications of trigonometry extends far beyond a one-week course. From geometry to circuit theory, sine and cosine appear in almost every branch of science and mathematics. I hope this book provided a tiny yet self-consistent introduction to the subject of trigonometry.

One more thing,

Before ending the session, remember to meditate and recall everything. If you are satisfied with the sessions, you are good to go. Have a good night sleep and comeback tomorrow for solving the 'Final Quiz'. If you are not satisfied. It's ok, tomorrow, return to the same session.

Don't give up.

Final Quiz

Even the hardest puzzles have a solution.

- Unknown

PART-A

1) Find the radian measures corresponding to the following degrees.

   i. $30°$ **ii.** $60°$ **iii.** $45°$ **iv.** $120°$

2) Find the degree measure corresponding to the following radians.

   i. $\dfrac{3\pi}{2}$ **ii.** 4π **iii.** 2π **iv.** $\dfrac{2\pi}{3}$

3) If $\sin\theta = \dfrac{1}{2}$ and θ is acute, find:

   i. $\cos\theta$ **ii.** $\tan\theta$ **iii.** $\sin 2\theta$

4) Find the values of:

   i. $\sin\left(\dfrac{\pi}{3}\right)\cos\left(\dfrac{\pi}{6}\right) + \cos\left(\dfrac{\pi}{3}\right)\cos\left(\dfrac{\pi}{6}\right)$

   ii. $\tan^2 60° + \tan^2 30°$

   iii. $4\sin^3\dfrac{\pi}{3} - 3\cos\dfrac{\pi}{6}$

5) Prove the following identities:

   i. $(1 + \cos\theta)(1 - \cos\theta) = \sin^2\theta$

   ii. $(\sin\theta + \cos\theta)^2 = 1 + 2\sin\theta\cos\theta$

   iii. $(\cot\theta - 1)^2 + (\cot\theta + 1)^2 = 2\csc^2\theta$

6) Find the values of:

   i. $\sin 120°$ **ii.** $\tan 180°$ **iii.** $\sin 210°$ **iv.** $\tan 330°$

PART-B

1) Find the radian measures corresponding to the following degrees:

   i. $47°25'36''$ **ii.** $52°30'24''$ **iii.** $64°24'38''$

   iv. $228°33'10''$

2) Find the degree measures corresponding to the following radians:

    i.   1rad   **ii.**   3rad   **iii.**   5rad   **iv.** $\dfrac{1969}{6750}$ rad

3) If $\sin\theta = -\dfrac{4}{5}$ and θ lies in the IV quadrant. Find all the other trigonometric functions.

4) Prove the following identities:

    i. $\dfrac{\cos\theta}{1+\sin\theta} = \dfrac{1-\sin\theta}{\cos\theta}$

    ii. $\sqrt{\dfrac{1-\sin\theta}{1+\sin\theta}} = \sec\theta - \tan\theta$

    iii. $\dfrac{1-\cos\theta}{1+\cos\theta} = (\operatorname{cosec}\theta - \cot\theta)^2$

5) If $\sin\theta = \dfrac{1}{2}$ and θ lies in the second quadrant, find:

    i. $\sin 2\theta$ **ii.** $\cos 2\theta$ **iii.** $\tan 2\theta$

6) Prove the following equations:

    i. $\sin 120° \cos 330° + \cos 240° \sin 330° = 1$

    ii. $\cos 570° \sin 510° - \sin 330° \cos 390° = 0$

PART-C

1) If $\cos A = -\dfrac{12}{13}$ and $\cot B = \dfrac{24}{7}$, where A lies in the second and B lies in the first quadrant, Find:

    i. $\sin(A + B)$ **ii.** $\cos(A - B)$ **iii.** $\tan(A + B)$

2) Show that:

    i. $\sin\left(\dfrac{\pi}{3} + \theta\right) - \sin\left(\dfrac{\pi}{3} - \theta\right) = \sin\theta$

    ii. $\sin(A + B)\sin(A - B) = \sin^2 A - \sin^2 B$

iii. $\dfrac{\sin(A+B)}{\sin(A-B)} = \dfrac{\tan A + \tan B}{\tan A - \tan B}$

3) Prove the following identities

i. $\dfrac{\sin 2\theta}{1+\cos 2\theta} = \tan\theta$

ii. $\dfrac{1-\cos 2\theta}{1+\cos 2\theta} = \tan^2\theta$

iii. $1 + \tan\theta \tan 2\theta = \sec 2\theta$

4) Prove the following identities:

i. $\dfrac{\sin 3\theta}{\sin\theta} + \dfrac{\cos 3\theta}{\cos\theta} = 4\cos 2\theta$

ii. $\dfrac{\cos 3\theta + \sin\theta}{\cos 3\theta - \cos\theta} = -\cot\theta$

iii. $\tan 3\theta = \dfrac{3\tan\theta - \tan^3\theta}{1 - 3\tan^2\theta}$

5) Show that:

i. $\dfrac{\cos\left(\frac{\pi}{2}+A\right)\sec(2\pi+A)\tan(\pi-A)}{\sec(A-4\pi)\sin(3\pi+A)\cot\left(A-\frac{\pi}{2}\right)} = 1$

ii. $\dfrac{\sin(180°+\theta)\cos(90°-\theta)\tan(270°+\theta)}{\sec(540°-\theta)\cos(360°+\theta)} = -\sin\theta\cos\theta$

6) Prove the following identities:

i. $\dfrac{\sin 2\theta + \sin 5\theta - \sin\theta}{\cos 2\theta + \cos 5\theta + \cos\theta} = \tan 2\theta$

ii. $\dfrac{\sin A + \sin 3A + \sin 5A}{\cos A + \cos 3A + \cos 5A} = \tan 3A$

NOTES

NOTES

* 9 7 8 9 3 5 5 2 6 8 8 9 1 *